Springer Topics in Signal Processing

Volume 6

For further volumes:
http://www.springer.com/series/8109

Jacob Benesty · Jingdong Chen

Study and Design of Differential Microphone Arrays

Springer

Prof. Jacob Benesty
INRS-EMT, University of Quebec
Montreal
QC, Canada

Prof. Jingdong Chen
Northwestern Polytechnical University
Xi'an
Shaanxi, China

ISSN 1866-2609 ISSN 1866-2617 (electronic)
ISBN 978-3-642-33752-9 ISBN 978-3-642-33753-6 (eBook)
DOI 10.1007/978-3-642-33753-6
Springer Heidelberg New York Dordrecht London

Library of Congress Control Number: 2012949066

Printed on acid-free paper

Springer is part of Springer Science+Business Media (www.springer.com)

Abstract

Microphone arrays have attracted a lot of interest over the last few decades since they have the potential to solve many important problems such as noise reduction/speech enhancement, source separation, dereverberation, spatial sound recording, and source localization/tracking, to name a few. However, the design and implementation of microphone arrays with beamforming algorithms is not a trivial task when it comes to processing broadband signals such as speech. Indeed, in most sensor arrangements, the beamformer tends to have a frequency-dependent response. One exception, perhaps, is the family of differential microphone arrays (DMAs) that have the promise to form frequency-independent responses. Moreover, they have the potential to attain high directional gains with small and compact apertures. As a result, this type of microphone arrays has drawn much research and development attention recently. This book is intended to provide a systematic study of DMAs from a signal processing perspective. The primary objective is to develop a rigorous but yet simple theory for the design, implementation, and performance analysis of DMAs. The theory includes some signal processing techniques for the design of commonly used first-order, second-order, third-order, and also the general Nth-order DMAs. For each order, particular examples are given on how to form standard directional patterns such as the dipole, cardioid, supercardioid, hypercardioid, subcardioid, and quadrupole. The study demonstrates the performance of the different order DMAs in terms of beampattern, directivity factor, white noise gain, and gain for point sources. The inherent relationship between differential processing and adaptive beamforming is discussed, which provides a better understanding of DMAs and why they can achieve high directional gain. Finally, we show how to design DMAs that can be robust against white noise amplification.

Contents

Chapter 1
Introduction

1.1 General Principles of Microphone Arrays

Ever since the invention of telephone systems, sound signal acquisition has become an essential part of acoustic and speech processing. Traditionally, sound signals are recorded and processed with the use of a single microphone; but this way of acquisition suffers significant limitations: not only the sound spatial realism is lost but also the flexibility in processing the recorded signal. Indeed, extracting the signal of interest from the noise, reverberation, and competing sources, is confined to a small family of techniques. To preserve the sound fidelity, spatial realism, and increase the processing flexibility in dealing with multiple sources and noise, the concept of microphone arrays has been introduced [1]–[5].

Simply put, a microphone array consists of a number of acoustic sensors, which are organized together to sample the sound field with spatial diversity. As a result, the array outputs contain the signal of interest, noise, interferences, and also the propagation information that is represented by the acoustic impulse responses from the radiating sources to the microphones. Through processing the array outputs, various functionalities can be implemented including but not limited to: localizing and tracking the sound sources, extracting the signal of interest, suppressing ambient noise, separating different sound sources, and recording spatial sound. The degree of how well these functionalities are accomplished, however, is conditioned on many factors such as the quality of each individual microphone, the number of sensors, the array geometry, the operating environment, and the processing algorithm (called beamforming). The difference can be enormous, depending on how the array is designed. Therefore, a considerable amount of attention has been paid to the design issues of microphone arrays and many different types of them have been developed over the last few decades. Broadly, those arrays can be categorized into two basic classes based on different principles: additive and differential arrays.

J. Benesty and J. Chen, *Study and Design of Differential Microphone Arrays*,
Springer Topics in Signal Processing 6, DOI: 10.1007/978-3-642-33753-6_1,

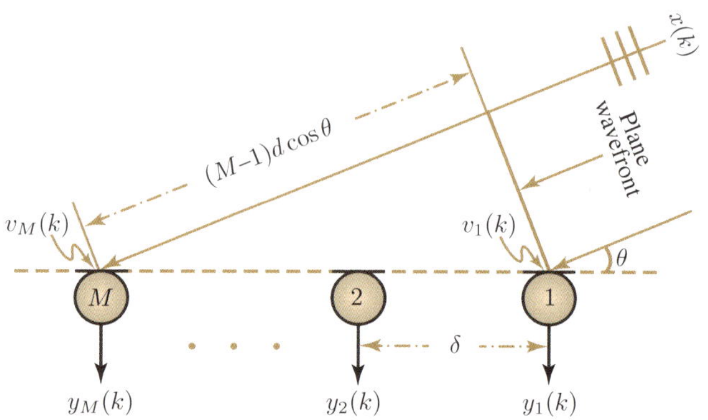

Fig. 1.1 Illustration of a uniformly spaced linear additive microphone array for sound capture in the farfield.

Additive arrays originally refer to the arrays that achieve signal enhancement and noise suppression based on the synchronize-and-add principles; but they have now evolved to include all the arrays with large inter-element spacing (from a couple of centimeters to a couple of decimeters) and optimal gain in broadside directions. In such arrays, each sensor converts the acoustic pressure from its own viewpoint into an electrical signal. By processing all the sensors' outputs, the signal of interest can be extracted while the unwanted noise and interference can be suppressed. To illustrate how an additive array works, let us consider a simple example with a uniformly spaced linear array of M microphones as shown in Fig. 1.1. Assume that there is a single desired source in the farfield such that its spherical wavefront appears planar at the array. If we neglect the propagation attenuation, the signal received at the discrete-time index k and at the mth microphone can be written as

$$\begin{aligned} y_m(k) &= x_m(k) + v_m(k) \\ &= x(k - t - \tau_m) + v_m(k),\ m = 1, 2, \ldots, M, \end{aligned} \tag{1.1}$$

where t is the propagation time from the source $x(k)$ to microphone 1 (reference point), τ_m is the relative time delay between the mth microphone and the first one, and $x_m(k) = x(k - t - \tau_m)$ and $v_m(k)$ are, respectively, the signal of interest and noise observed at the mth microphone. The noise signal, $v_m(k)$, is assumed to be uncorrelated with $x(k)$. With a uniform linear array and a farfield source, the delay τ_m can be expressed into the following form according to the geometry shown in Fig. 1.1:

$$\tau_m = \frac{(m-1)\delta\cos\theta}{c},\ m = 1, 2, \ldots, M, \tag{1.2}$$

where δ is the spacing between two neighboring microphones, c is the speed of sound in the air, and θ is the signal incident angle. For simplicity of exposition, let us work in the frequency domain. Then, the signal model in (1.1) becomes

$$\begin{aligned} Y_m(\omega) &= X_m(\omega) + V_m(\omega) \\ &= X(\omega)e^{-\jmath\omega\,(t+\tau_m)} + V_m(\omega),\ m = 1, 2, \ldots, M, \end{aligned} \tag{1.3}$$

where $\omega = 2\pi f$ is the angular frequency, f denotes the temporal frequency, $\jmath = \sqrt{-1}$ represents the imaginary unit, and $Y_m(\omega)$, $X_m(\omega)$, $V_m(\omega)$, and $X(\omega)$ are the frequency-domain representations of $y_m(k)$, $x_m(k)$, $v_m(k)$, and $x(k)$, respectively.

Now, let us process the M signals $Y_m(\omega)$, $m = 1, 2, \ldots, M$, in order to extract the desired signal $X(\omega)$ (up to a delay) and reduce the effect of $V_m(\omega)$. The most straightforward and simple way of doing this is through the use of the so-called delay-and-sum (DS) beamformer. The basic principle of this approach is that it compensates $y_m(k)$ with a delay τ_m to align all the M microphone signals, i.e., multiply $Y_m(\omega)$ with $e^{\jmath\omega\tau_m}$, and then average the results together. The DS beamformer output is then

$$\begin{aligned} Z(\omega) &= \frac{1}{M}\sum_{m=1}^{M} Y_m(\omega)e^{\jmath\omega\tau_m} \\ &= X(\omega)e^{-\jmath\omega t} + \frac{1}{M}\sum_{m=1}^{M} V_m(\omega)e^{\jmath\omega\tau_m}. \end{aligned} \tag{1.4}$$

To check whether the beamformer output is less noisy than its input, let us compare the input and output signal-to-noise ratios (SNRs). The input SNR of the DS beamformer, according to the signal model given in (1.3), is defined as the SNR at the first (reference) microphone, i.e.,

$$\begin{aligned} \mathrm{iSNR}(\omega) &= \frac{\phi_{X_1}(\omega)}{\phi_{V_1}(\omega)} \\ &= \frac{\phi_X(\omega)}{\phi_{V_1}(\omega)}, \end{aligned} \tag{1.5}$$

where $\phi_{X_1}(\omega) = E\left[|X_1(\omega)|^2\right]$, $\phi_{V_1}(\omega) = E\left[|V_1(\omega)|^2\right]$, and $\phi_X(\omega) = E\left[|X(\omega)|^2\right]$, with $E[\cdot]$ denoting mathematical expectation. The output SNR of the beamformer, from (1.4), is then

$$\mathrm{oSNR}(\omega) = \frac{\phi_X(\omega)}{\frac{1}{M^2}E\left[\left|\sum_{m=1}^{M} V_m(\omega)e^{\jmath\omega\tau_m}\right|^2\right]}. \tag{1.6}$$

Now, we evaluate the following two cases.

- The noise signals at the different microphones are uncorrelated.

If all the noise signals, $V_m(\omega)$, $m = 1, 2, \ldots, M$, are uncorrelated with each other and have the same variance, it is easy to check that the gain in SNR is

$$\mathcal{G}(\omega) = \frac{\mathrm{oSNR}(\omega)}{\mathrm{iSNR}(\omega)} = M. \tag{1.7}$$

In this scenario, we see that a simple DS beamformer can improve the SNR by a factor of M or $10 \log_{10} M$ in dB and the gain in SNR is independent of frequency.

- The noise signals are from a point source with an incident angle of θ_{n}. In this case, we have

$$V_m(\omega) = V_1(\omega) e^{-\jmath\omega(m-1)\delta\cos\theta_{\mathrm{n}}/c}. \tag{1.8}$$

Substituting $V_m(\omega)$ into (1.6) and with some mathematical manipulations, one can have

$$\mathrm{oSNR}(\omega) = \mathrm{iSNR}(\omega) \left| \frac{M \sin\left[\omega\delta(\cos\theta - \cos\theta_{\mathrm{n}})/(2c)\right]}{\sin\left[M\omega\delta(\cos\theta - \cos\theta_{\mathrm{n}})/(2c)\right]} \right|^2. \tag{1.9}$$

Therefore, the gain in SNR is

$$\mathcal{G}(\omega) = \left| \frac{M \sin\left[\omega\delta(\cos\theta - \cos\theta_{\mathrm{n}})/(2c)\right]}{\sin\left[M\omega\delta(\cos\theta - \cos\theta_{\mathrm{n}})/(2c)\right]} \right|^2. \tag{1.10}$$

It is seen that $\mathcal{G}(\omega)$ is a function of the number of sensors, the microphone spacing, the angular separation between the signal of interest and the noise signal to be suppressed, and frequency. Figure 1.2 plots the gain in SNR with a uniform linear array of ten microphones.

Another way to study the performance of an additive array is through the directivity pattern (also called beampattern). For the signals given in (1.3), if we neglect the noise terms and compensate the mth microphone signal with a delay equal to $(m-1)\delta\cos\theta_{\mathrm{s}}/c$, where θ_{s} is the angle of the desired source signal, the DS beamformer output is

$$Z(\omega) = \frac{1}{M} \sum_{m=1}^{M} X(\omega) e^{-\jmath\omega t} e^{-\jmath\omega(m-1)\delta(\cos\theta - \cos\theta_{\mathrm{s}})/c}. \tag{1.11}$$

The beampattern, which is defined as the magnitude of the transfer function between the beamformer output and the input signal, is then

$$\mathcal{B}(\omega, \theta) = \left| \frac{Z(\omega)}{X(\omega)} \right| = \left| \frac{\sin\left[M\omega\delta(\cos\theta - \cos\theta_{\mathrm{s}})/(2c)\right]}{M \sin\left[\omega\delta(\cos\theta - \cos\theta_{\mathrm{s}})/(2c)\right]} \right|. \tag{1.12}$$

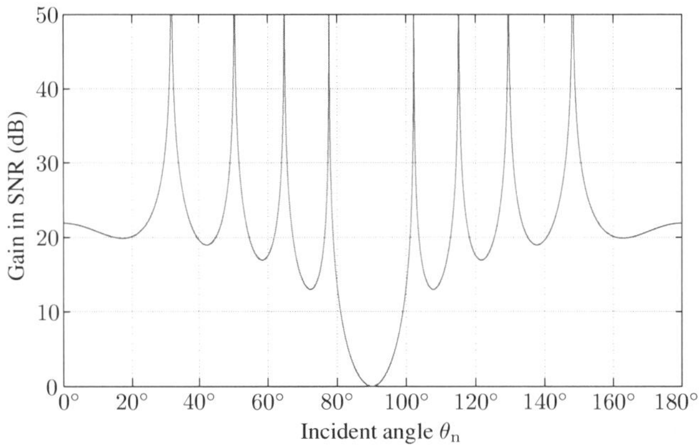

Fig. 1.2 Gain in SNR of the DS beamformer with a uniformly spaced linear array for a point noise source with incident angle θ_n. $M = 10$, $\theta = 90°$, $\delta = 8$ cm, and $f = 2$ kHz.

Figure 1.3 plots the beampattern when $M = 10$, $\delta = 8$ cm, $\theta_s = 90°$, and $f = 2$ kHz. It consists of a total of nine beams[1]. The one with the highest amplitude is called the main lobe and all the others are called side lobes. One important parameter regarding the main lobe is the so-called beamwidth (or main lobe width), which is defined as the region between the first zero-crossings on either side of the main lobe. For a DS beamformer with a linear array, the beamwidth is $2\cos^{-1}[c/(Mf\delta)]$. The beampattern of Fig. 1.3 indicates that the DS beamformer allows the desired signal from the look direction (i.e., $\theta = \theta_s$) to pass through without attenuation, while suppressing noise and other interfering signals coming from other directions. The degree of suppression depends on the number of sensors, the microphone spacing, the angular separation between the desired signal and the signals to be suppressed, and frequency.

The above simple DS beamformer, though illustrated with a linear array, can be used in any geometry of arrays such as the circular, planar, and spherical ones. However, the use of a DS beamformer to process broadband speech signals suffers from the following number of problems and drawbacks. 1) Its beampattern is frequency dependent and the beamwidth is inversely proportional to the frequency. Therefore, this beamformer is not effective in dealing with low-frequency noise and interference. 2) Noise is not uniformly attenuated over its entire spectrum, resulting in some disturbing artifacts in the array output [6]. 3) If the incident angle of the speech source is different from the array look direction, even slightly, the speech signal will be low-pass filtered, leading to speech distortion. To overcome these drawbacks, broadband beamforming techniques have been developed with the subband and filter-and-sum frameworks [7], [8]. In the former structure, the array signals

[1] For a properly designed array, the number of beams in the range between 0° and 180° is equal to $M - 1$.

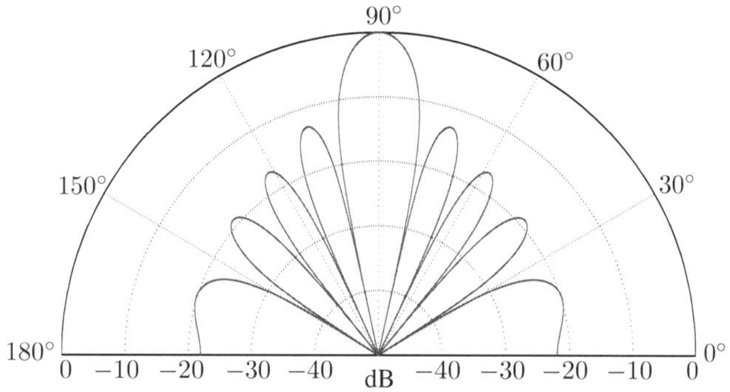

Fig. 1.3 Beampattern of the DS beamformer with a uniformly spaced linear array. $M = 10$, $\theta_s = 90°$, $\delta = 8$ cm, and $f = 2$ kHz.

are decomposed into multiple subbands. A narrowband beamformer is designed in each subband with a constraint applied to control the beamwidth so that all the beamformers from different subbands have the same beamwidth. Though it can make constant beamwidth across a wide range of frequencies, this way of broadband beamforming sacrifices the array performance in high frequencies. In comparison, the filter-and-sum structure applies a finite-impulse-response (FIR) filter to each sensor signal and then sums up all the filtered signals to form the array output. The core problem in this broadband beamforming is to determine the coefficients of those FIR filters, which can be accomplished either independently of the acoustic environment and array data (fixed beamformer) or in an adaptive manner according to the received array data such as the minimum variance distortionless (MVDR) and linearly constrained minimum variance (LCMV) algorithms [7], [9]–[14]. Generally speaking, adaptive beamformers can be more efficient than the fixed ones in suppressing reverberation and competing sources; but they may suffer from desired signal cancellation, which deserves careful attention.

Additive arrays have been intensively studied in the literature. For the interested reader, see [2], [4], [5] for a more comprehensive coverage of such arrays and their processing algorithms. In general, additive arrays are large in size and they may be effective in processing high frequency signals while they may not be at low frequencies.

Differential microphone arrays (DMAs) refer to the arrays that are responsive to the spatial derivatives of the acoustic pressure field. The basic idea of such arrays can be traced back to the 1930s when the directional ribbon microphones were invented [15], [16]. The earliest such directional microphones were constructed by combining omnidirectional pressure sensors and gradient ribbon sensors[2] together to produce the desired directional patterns such as the dipole, cardioid, hypercardioid, and supercardioid. This idea was then

[2] Note that ribbon microphones are gradient sensors in nature because they directly measure the sound pressure gradient and not the sound pressure itself.

extended to the directional electret (capacitor) microphones [17], [18]. Unlike the ribbon microphones, electrets are pressure sensors and they have an omnidirectional response. To form a directional pattern, one would need to measure the differentials of the acoustic pressure field, which can be achieved by combining the outputs of a number of omnidirectional sensors. For instance, the first-order differential of the acoustic pressure can be obtained by subtracting two closely spaced omnidirectional microphones' outputs while a general Nth-order differential of the acoustic pressure can be formed by subtracting two differentials of order $N-1$. A directional microphone consists of two or more omnidirectional microphones and, therefore, it is indeed a microphone array. (Note that today, a directional microphone may have only a single transducing element and the desired pattern is achieved by using calculated front-back delay paths [19].) However, it is named a microphone instead of a microphone array in its early form primarily because all the sensors inside such a system are physically conjoined in a single housing so the overall system still looks like a single microphone.

Directional microphones, in their early form, have a prominent limitation, i.e., once made, their directional response is fixed. If, for a given application, we find that the microphone we bought does not produce the expected performance, the only option is to try a different type. There is no flexibility in adapting the directional pattern to fit the application needs. To circumvent this limitation, the modern concept of DMAs was developed, in which a number of pressure microphones are arranged into a particular geometry and digital signal processing techniques are then used to process the microphones' outputs to obtain the desired directional response [20]–[35]. Figure 1.4 illustrates how first-, second-, and third-order DMAs are constructed with a linear geometry. Basically, a general Nth-order DMA has a response proportional to a linear combination of signals derived from spatial derivatives from order 0 to (including) order N. Note that an inherent assumption in the construction of a DMA is that the microphones are placed close enough so that the true acoustic pressure differentials can be approximated by finite differences between microphone sensors' outputs. This is one of the major reasons why a DMA is in general compact in size.

Besides its small size, a DMA has the following number of advantages in comparison with an additive array. 1) It can form frequency-invariant beampatten and, therefore, it is more suitable to process broadband speech signals. 2) It is effective not only for high frequencies, but for low frequencies as well. 3) For a given number of sensors, differential arrays have the potential to attain maximum directional gain [15].

Although a DMA has many attractive properties, its design and implementation is by no means a trivial task. First, the response of an Nth order array has a high-pass filter nature with a slope of $6N$ dB/octave, so its frequency response has to be properly compensated to process broadband speech signals. Second, the frequency response and level of a DMA are sensitive to the position and orientation of the arrays relative to the sound source. Con-

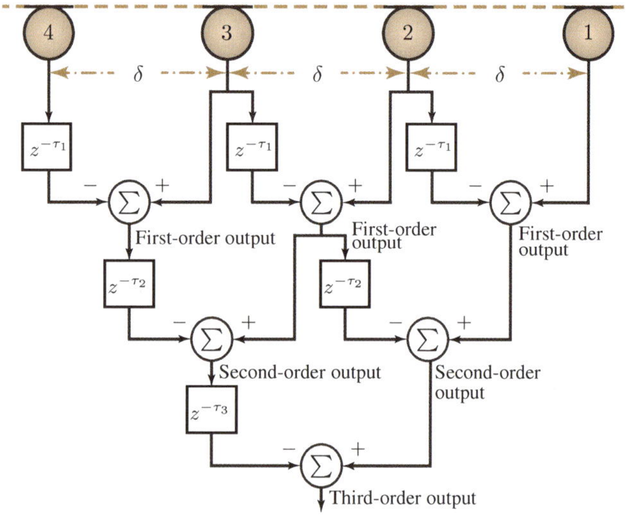

Fig. 1.4 A schematic diagram of construction of first-, second-, and third-order DMAs.

sequently, it is necessary to perform frequency and level equalization to its response according to the range and incident angle of the sound source [16]. However, this equalization process, though theoretically feasible, is in general very difficult for a DMA with an order higher than two. Finally, a DMA suffers greatly from white noise amplification, particularly at low frequencies. Although some efforts in the literature have been devoted to solving these issues, they have not been fully addressed.

In summary, DMAs have the great potential to solve many important problems in voice communication. However, unlike the additive arrays that have been intensively studied in the past, differential arrays are less addressed so far. Many aspects of DMAs are still unclear if not a mystery. This book is, therefore, devoted to providing a systematic study of DMAs from a signal processing perspective. The primary objective is to develop fundamental theory and algorithms associated with the design and implementation of DMAs as well as principles to analyze and explain their performance and limitations. Coverage of the theory includes signal processing algorithms for the design of commonly used first-order, second-order, third-order, as well as the general Nth-order DMAs. For each order of DMAs, particular examples are given on how to form standard directivity patterns such as the dipole, cardioid, supercardioid, hypercardioid, subcardioid, and quadrupole. Study is also provided to show the performance of different order DMAs and the related beamforming algorithms in terms of beampattern, directivity factor, white noise gain, and gain for a point source. The inherent relationship between differential beamforming and adaptive beamforming is discussed, which provides a dif-

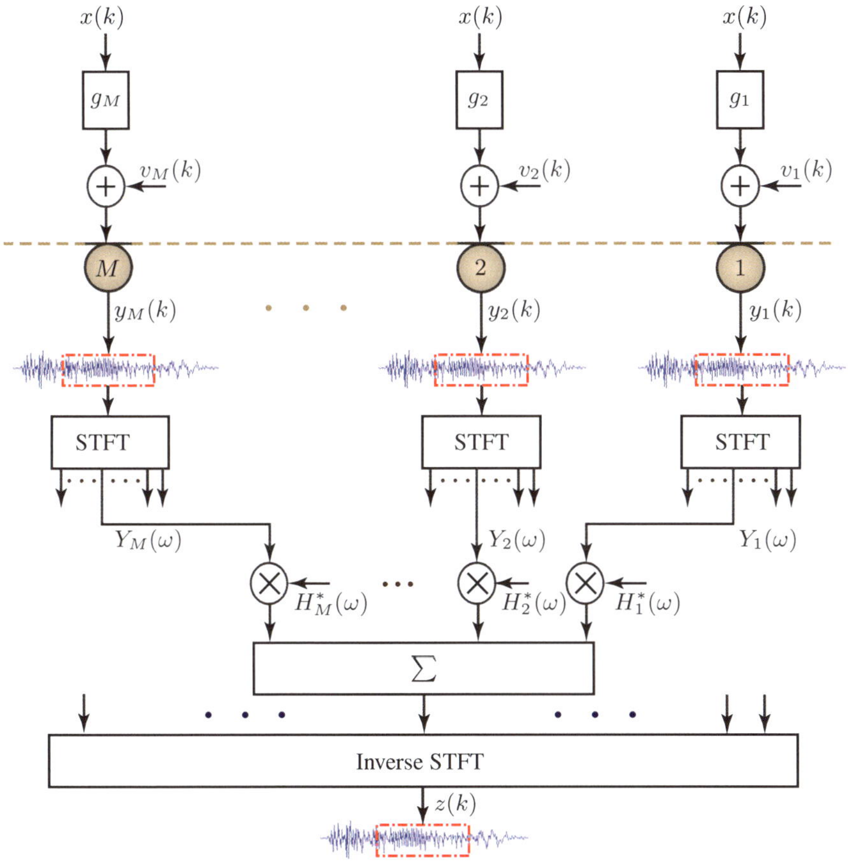

Fig. 1.5 General block diagram of a DMA system for processing speech signals.

ferent way to gain a better understanding why DMAs can achieve maximal directional gain.

1.2 The Paradigm for the Design and Implementation of Differential Arrays

This book is designed to address the challenging problem of extracting a desired speech signal from interference and noise using DMA techniques. Different from the traditional approach that constructs a general Nth order DMA in the time domain as illustrated in Fig. 1.4, we adopt a new and more general paradigm, in most of this work, as shown in Fig. 1.5. This paradigm starts with a clean speech signal $x(k)$ (from a desired speaker) passing through some acoustic environment and then being corrupted by some unwanted noise. A DMA consisting of M microphones is used to pick up the signals. To estimate the source signal, the M noisy signals are partitioned into small overlapping frames of a few milliseconds. Each frame is transformed into the short-time Fourier transform (STFT) domain. In each subband, a differential beamformer is designed and applied to the multichannel signals, thereby producing an estimate of the clean signal spectrum in this subband. Finally, the time-domain clean speech estimate is constructed using the overlap-add or overlap-save technique with the inverse STFT. Apparently, the most crucial step in this paradigm is the design of the differential beamformers, which is one of the main focuses of this book.

The problem addressed in this work is of great importance from both the theoretical and practical viewpoints. From the theoretical side, it is essential to understand the theory and principles underlying DMAs from a signal processing perspective. On the practical side, DMAs can be used in a wide range of applications such as teleconferencing, hands-free communications, 3D sound recording, mobile phones, hearing aids, bluetooth headphones, and navigation systems in cars, just to name a few. Therefore, it is important to understand the design issues of DMAs and their limitations as well as the way to circumvent those limitations.

1.3 Organization of the Book

The material in this book is organized into seven chapters, including this one. In the next six chapters, we attempt to cover the most basic concepts and fundamental techniques used in the design and implementation of the different orders of DMAs and the associated beamforming algorithms. All is explained from a signal processing perspective. The material discussed in these chapters is as follows.

Chapter 2 presents the general formulation of a DMA design problem. It also introduces several basic concepts that are important and useful in the design and evaluation of a DMA including the steering vector (for a plane wave with the conventional anechoic and farfield model), the beampattern, the gain in SNR, the white noise gain, the directivity factor (the SNR gain in diffuse noise), etc. In the design of a DMA, a particularly structured matrix called the Vandermonde matrix generally appears (explicitly or implicitly) in the formulation. This matrix and its inverse is also discussed.

The simplest form of a DMA is the first-order one that uses only two omnidirectional sensors. If the two microphones are properly placed, many beampatterns, such as the dipole, cardioid, subcardioid, hypercardioid, and supercardioid, can be formed by processing the two microphones' outputs. Chapter 3 is dedicated to the design of first-order DMAs. It discusses how to form different beampatterns and studies the performance associated with each beampattern. It also illustrates the impact of the sensor spacing on the performance.

Chapter 4 is concerned with the design and implementation of second-order DMAs. It studies how to form the popularly used beampatterns such as the dipole, two types of cardioid, quadrupole, hypercardioid, and supercardioid. The performance of these beamformers in terms of the SNR gain in spatially correlated and uncorrelated noise is also presented.

Chapter 5 investigates a family of third-order DMAs. Their beampatterns have three distinct nulls, which can be viewed as three fundamental constraints. By properly adjusting these constraints, different beampatterns can be formed, leading to different performance in suppressing noise impinging at the array from different directions. As examples, we show how to design the hypercardioid and supercardioid beampatterns and study their SNR gain in white, diffuse, and point source noise. We also generalize this method to the design of Nth order DMAs with N distinct nulls.

The performance of a DMA in reducing noise depends on several factors such as the order of the DMA and the characteristics of the noise. In general, the higher the order of the DMA, the more effective is the array in suppressing noise. However, a high-order DMA leads to huge white noise amplification, which is a significant barrier that prevents using them. In Chapter 6, we show how DMAs are related to the classical adaptive beamforming technique. From this important and useful relationship, we then derive a minimum-norm solution for the design of a general Nth-order DMA. Unlike the traditional way of DMA design where exactly $N + 1$ sensors are required to design an Nth order DMA, the proposed approach can use more than $N + 1$ sensors to design an Nth order DMA. The advantage of this technique is its robustness against white noise amplification. The more the number of microphones, the more robust is the DMA.

In Chapter 7, we show that the patterns of DMAs can be obtained from the general definition of the beampattern by approximating the exponential function with its MacLaurin's series expansion. In other words, a directional pattern of order N can be obtained from the MacLaurin's series of order N, as long as this approximation holds. In this chapter, we also show how to design differential arrays based on this approach and their relationship to adaptive beamforming.

References

1. J. L. Flanagan, J. D. Johnson, R. Zahn, and G. W. Elko, "Computer-steered microphone arrays for sound transduction in large rooms," *J. Acoust. Soc. Amer.*, vol. 75, pp. 1508-1518, Nov. 1985.
2. M. Branstein and D. B. Ward, Eds., *Microphone Arrays: Signal Processing Techniques and Applications*. Berlin, Germany: Springer-Verlag, 2001.

3. J. Benesty and Y. Huang, Eds., *Adaptive Signal Processing: Applications to Real-World Problems*. Berlin, Germany: Springer-Verlag, 2003.
4. Y. Huang, J. Benesty, and J. Chen, *Acoustic MIMO Signal Processing*. Berlin, Germany: Springer-Verlag, 2006.
5. J. Benesty, J. Chen, and Y. Huang, *Microphone Array Signal Processing*. Berlin, Germany: Springer-Verlag, 2008.
6. D. B. Ward, R. C. Williamson, and R. A. Kennedy, "Broadband microphone arrays for speech acquisition," *Acoust. Australia*, vol. 26, pp. 17-20, Apr. 1998.
7. O. L. Frost, III, "An algorithm for linearly constrained adaptive array processing," *Proc. IEEE*, vol. 60, pp. 926–935, Aug. 1972.
8. J. Benesty, J. Chen, Y. Huang, and J. Dmochowski, "On microphone-array beamforming from a MIMO acoustic signal processing perspective," *IEEE Trans. Audio, Speech, Language Process.*, vol. 15, pp. 1053–1065, Mar. 2007.
9. B. Widrow, P. Mantey, L. Griffiths, and B. Goode, "Adaptive antenna systems," *Proc. IEEE*, vol. 55, pp. 2143–2159, Dec. 1967.
10. J. Capon, "High resolution frequency-wavenumber spectrum analysis," *Proc. IEEE*, vol. 57, pp. 1408–1418, Aug. 1969.
11. W. F. Gabriel, "Adaptive arrays–an introduction," *Proc. IEEE*, vol. 64, pp. 239–272, Feb. 1976.
12. R. A. Monzingo and T. W. Miller, *Introduction to Adaptive Arrays*. New York, NY: Wiley, 1980.
13. T. J. Shan and T. Kailath, "Adaptive beamforming for coherence signals and interference," *IEEE Trans. Acoust., Speech, Signal Process.*, vol. ASSP-33, pp. 527–536, 1985.
14. H. Cox, R. M. Zeskind, and M. M. Owen, "Robust adaptive beamforming," *IEEE Trans. Acoust., Speech, Signal Process.*, vol. ASSP-35, pp. 1365–1376, Oct. 1987.
15. H. F. Olson, "A uni-directional ribbon microphone," *J. Acoust. Soc. Am.*, vol. 3, pp. 315, 1932.
16. H. F. Olson, "Gradient microphones," *J. Acoust. Soc. Am.*, vol. 17, pp. 192–198, Jan. 1946.
17. G. M. Sessler and J. E. West, "Directional transducers," *IEEE Trans. Audio Electroacoustic.*, vol. 19, pp. 19–23, Mar. 1971.
18. P. Graven and M. Gerzon, "Coincident microphone simlation covering three dimensional space and yielding various directional outputs," U.S. Patent 4,042,779, Aug. 1997.
19. J. Eargle, *The Microphone Book*. Burlington, MA: Focal Press, 2005.
20. G. W. Elko, "Steerable and variable first order differential microphone array," U.S. Patent 6,041,127, Apr. 1997.
21. G. W. Elko and A.-T. N. Pong, "A steerable and variable first-order differential microphone array," in *Proc. IEEE ICASSP*, 1997, pp. 223–226.
22. G. W. Elko, "Superdirectional microphone arrays," in *Acoustic Signal Processing for Telecommunication*, S. L. Gay and J. Benesty, Eds. Boston, MA: Kluwer Academic Publishers, 2000, Chapter 10, pp. 181–237.
23. G. W. Elko and J. Meyer, "Microphone arrays," in *Springer Handbook of Speech Processing*, J. Benesty, M. M. Sondhi, and Y. Huang, Eds. Berlin: Germany, Springer-Verlag, 2008, Chapter 50, Part I, pp. 1021–1041.
24. M. Williamsen, "Design and measurement of a dipole microphone," *AudioXpress*, vol. 7, pp. 1–8, 2009.
25. H. Teutsch and G. W. Elko, "First- and second-order adaptive differential microphone arrays," in *Proc. IWAENC*, 2001.
26. J. Pekonen, "Microphone techniques for spatial sound," in *Proc. Acoustics Seminar on Spatial Sound Modeling*, 2008.
27. T. D. Abhayapala and A. Gupta, "Higher order differential-integral microphone arrays," *J. Acoust. Soc. Am.*, vol. 127, pp. EL227–EL233, May 2010.

28. M. Buck and M. Rößler, "First order differential microphone arrays for automotive applications," in *Proc. IWAENC*, 2001.
29. J. Bitzer, K.-D. Kammeyer, and K. U. Simmer, "An alternative implementation of the superdirective beamformer," in *Proc. IEEE WASPAA*, 1999.
30. M. Buck, "Aspects of first-order differential microphone arrays in the presence of sensor imperfections," *European Trans. Telecommunications*, vol. 13, pp. 115–122, Mar.-Apr. 2002.
31. E. De Sena, H. Hacihabiboğlu, and Z. Cvetković, "On the design and implementation of higher-order differential microphones," *IEEE Trans. Audio, Speech, Language Process.*, vol. 20, pp. 162–174, Jan. 2012.
32. V. Hamacher, J. Chalupper, J. Eggers, E. Fischer, U. Kornagel, H. Puder, and U. Rass, "Signal processing in high-end hearing aids: state of the art, challenges, and future trends," *EURASIP J. Applied Signal Process.*, vol. 2005, pp. 2915–2929, Jan. 2005.
33. R. Derkx, "Optimal azimuthal steering of a first-order supperdirectional microphone response," in *Proc. IWAENC*, 2008.
34. R. N. Miles, Y. Liu, Q. Su, and E. Cui, "A silicon directional microphone with second-order directivity," in *Proc. inter. Congress Acoust.*, 2007.
35. M. Kolundžija, C. Faller, and M. Vetterli, "Spatiotemporal gradient analysis of differential microphone arrays," *J. Audio Eng. Soc.*, vol. 59, pp. 20–28, Jan.-Feb. 2011.

Chapter 2
Problem Formulation

In this chapter, we explain some important aspects of beamforming and differential arrays. The problem of a DMA design is formulated while we progress in defining some useful concepts. We start with the definition of the steering vector for a plane wave with the conventional anechoic farfield model. We give the general definition of the beampattern as well as its expression for directional arrays. We then derive the gain in signal-to-noise ratio (SNR), which can be very useful in the evaluation of DMAs under different types of noise. Finally, we discuss the Vandermonde matrix, which always appears, explicitly or implicitly, in the design of DMAs.

2.1 Signal Model

We consider a source signal (plane wave) that propagates in an acoustic environment (anechoic farfield model) at the speed of sound, i.e., $c = 340$ m/s, and impinges on a uniform linear sensor array consisting of M omnidirectional microphones, where the distance between two successive sensors is equal to δ (see Fig. 2.1). The direction of the source signal to the array is parameterized by the angle θ. In this scenario, the corresponding steering vector (of length M) is

$$\begin{aligned} \mathbf{d}(\omega, \cos\theta) &= \begin{bmatrix} 1 \; e^{-\jmath\omega\delta\cos\theta/c} \; \cdots \; e^{-\jmath(M-1)\omega\delta\cos\theta/c} \end{bmatrix}^T \\ &= \begin{bmatrix} 1 \; \left(e^{-\jmath\omega\tau_0\cos\theta}\right)^1 \; \cdots \; \left(e^{-\jmath\omega\tau_0\cos\theta}\right)^{M-1} \end{bmatrix}^T, \end{aligned} \tag{2.1}$$

where the superscript ^{T} is the transpose operator, $\jmath = \sqrt{-1}$ is the imaginary unit, $\omega = 2\pi f$ is the angular frequency, $f > 0$ is the temporal frequency, and $\tau_0 = \delta/c$ is the delay between two successive sensors at the angle $\theta = 0°$. The acoustic wavelength is $\lambda = c/f$. In DMAs [1], it is always assumed that

J. Benesty and J. Chen, *Study and Design of Differential Microphone Arrays*,
Springer Topics in Signal Processing 6, DOI: 10.1007/978-3-642-33753-6_2,

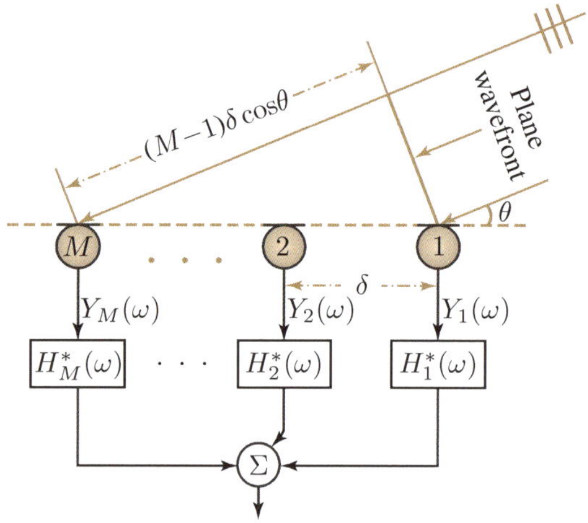

Fig. 2.1 A uniform linear microphone array with processing.

the sensor spacing, δ, is much smaller than the acoustic wavelength, λ, i.e., $\delta \ll \lambda$, implying that

$$\frac{\omega\delta}{c} = \omega\tau_0 \ll 2\pi. \tag{2.2}$$

The condition (2.2) easily holds for small values of δ and at low frequencies but not at high frequencies. With this condition, spatial aliasing, which has the negative effect of creating grating lobes (i.e., copies of the main lobe, which usually points toward the desired signal), is also avoided [2].

We consider fixed directional beamformers[1], like in DMAs, where the main lobe is at the angle $\theta = 0°$ (endfire direction) and the desired signal propagates at the same angle. This position is optimal as will become clearer later. Electronic steering (in the sense that the main lobe can be oriented to any possible direction without affecting the shape of the beampattern) with a uniform linear DMA is not really feasible but we will study some simple possibilities.

As pointed out in [3], there is a fundamental difference between differential arrays and filter-and-sum beamformers. In the latter category, the filters are optimized in such a way that the microphone signals are aligned in order to steer the main lobe in the direction of the desired signal, whereas in the former category the gains are optimized to steer a number of nulls in some specific directions.

The focus of this work is on the design, with small apertures, of beamformers whose beampatterns are very close to the ones obtained with "ideal"

[1] The terms beamformer, beamforming, and beampattern may not be adequate in the context of DMAs but we will still use them for convenience.

DMAs. For that, a complex weight, $H_m^*(\omega)$, $m = 1, 2, \ldots, M$, is applied at the output of each microphone, where the superscript * denotes complex conjugation. The weighted outputs are then summed together to form the beamformer output as shown in Fig. 2.1. Putting all the gains together in a vector of length M, we get

$$\mathbf{h}(\omega) = \left[H_1(\omega)\ H_2(\omega) \cdots H_M(\omega) \right]^T . \tag{2.3}$$

The objective then is to design such a filter for any directivity pattern of any order. The approach taken here is based on the fundamental observation that for all beampatterns of interest, some constraints must be fulfilled at all frequencies given that the number of microphones is equal to M. In other words, we select M fundamental constraints from a well-defined beampattern of a DMA. For example, in the first-order dipole with two microphones, the two fundamental constraints are a one at the angle $0°$ and a null at the angle $90°$. Since we have two microphones and two constraints, we have a simple linear system of two equations to solve. As a result, the obtained solution is optimal from a mathematical point of view and the derived dipole is the best we can get.

In the next two sections, we discuss some fundamental measures. We are only interested in narrowband measures. The broadband measures can be easily deduced from their respective narrowband counterparts.

2.2 Beampattern

Each beamformer has a pattern of directional sensitivity, i.e., it has different sensitivities from sounds arriving from different directions. The beampattern or directivity pattern describes the sensitivity of the beamformer to a plane wave (source signal) impinging on the array from the direction θ. Mathematically, it is defined as

$$\begin{aligned} \mathcal{B}\left[\mathbf{h}(\omega), \theta\right] &= \mathbf{d}^H(\omega, \cos\theta)\, \mathbf{h}(\omega) \\ &= \sum_{m=1}^{M} H_m(\omega)\, e^{\jmath(m-1)\omega\tau_0 \cos\theta}, \end{aligned} \tag{2.4}$$

where the superscript ^{H} is the transpose-conjugate operator.

The frequency-independent beampattern of an Nth-order DMA is well known. It is defined as [4]

$$\mathcal{B}_N(\theta) = \sum_{n=0}^{N} a_{N,n} \cos^n \theta, \tag{2.5}$$

where $a_{N,n}$, $n = 0, 1, \ldots, N$, are real coefficients. The different values of these coefficients determine the different directional patterns of the Nth-order DMA. In the direction of the desired signal, i.e., for $\theta = 0°$, the beampattern must be equal to 1, i.e., $\mathcal{B}_N(0°) = 1$. Therefore, we have

$$\sum_{n=0}^{N} a_{N,n} = 1. \tag{2.6}$$

As a result, we always choose the first coefficient as

$$a_{N,0} = 1 - \sum_{n=1}^{N} a_{N,n}. \tag{2.7}$$

It follows from (2.5) that an Nth-order DMA has at most N (distinct) nulls. All interesting patterns have at least one null in some direction. Since $\cos\theta$ is an even function, so is $\mathcal{B}_N(\theta)$. Therefore, on a polar plot, $\mathcal{B}_N(\theta)$ is symmetric about the axis $0° - 180°$ and any DMA design can be restricted to this range. Polar patterns are a very convenient way to describe the directional sensitivity of the DMAs.

The directivity factor (see also Section 2.3) of an Nth-order DMA, defined as the ratio between the directivity pattern at the endfire direction $\theta = 0°$ and the averaged directivity pattern over the whole space, is[2] [4], [5], [6]

$$\begin{aligned} \mathcal{G}_N &= \frac{\mathcal{B}_N^2(0°)}{\dfrac{1}{\pi}\displaystyle\int_0^{\pi} \mathcal{B}_N^2(\theta)\, d\theta} \\ &= \frac{\pi}{\displaystyle\int_0^{\pi} \left(\sum_{n=0}^{N} a_{N,n} \cos^n \theta \right)^2 d\theta} \end{aligned} \tag{2.8}$$

and what we call the directivity index is

$$\mathcal{D}_N = 10 \log_{10} \mathcal{G}_N. \tag{2.9}$$

We find that the first-order, second-order, and third-order directivity factors are

$$\mathcal{G}_1 = \frac{1}{a_{1,0}^2 + \dfrac{1}{2} a_{1,1}^2}, \tag{2.10}$$

$$\mathcal{G}_2 = \frac{1}{a_{2,0}^2 + \dfrac{1}{2} a_{2,1}^2 + \dfrac{3}{8} a_{2,2}^2 + a_{2,0} a_{2,2}}, \tag{2.11}$$

[2] This situation corresponds to the cylindrically isotropic noise field.

$$\mathcal{G}_3 = \frac{1}{a_{3,0}^2 + \frac{1}{2}a_{3,1}^2 + \frac{3}{8}a_{3,2}^2 + a_{3,0}a_{3,2} + \frac{5}{16}a_{3,3}^2 + \frac{3}{4}a_{3,1}a_{3,3}}. \quad (2.12)$$

The hypercardioid is the pattern obtained from the maximization of the directivity factor[3].

The front-to-back ratio is defined as the ratio of the power of the output of the array to signals propagating from the front-half plane to the output power for signals arriving from the rear-half plane [7]. This ratio, for the cylindrically isotropic noise field, is mathematically defined as [4], [7]

$$\mathcal{F}_N = \frac{\int_0^{\pi/2} \mathcal{B}_N^2(\theta)\, d\theta}{\int_{\pi/2}^{\pi} \mathcal{B}_N^2(\theta)\, d\theta}. \quad (2.13)$$

The supercardioid is the pattern obtained from the maximization of the front-to-back ratio[4] [7].

First-order directivity patterns have the form:

$$\mathcal{B}_1(\theta) = (1 - a_{1,1}) + a_{1,1}\cos\theta \quad (2.14)$$

and the most important ones are as follows.

- Dipole: $a_{1,1} = 1$, null at $\cos\theta = 0$, and $\mathcal{D}_1 = 3$ dB.
- Cardioid: $a_{1,1} = \frac{1}{2}$, null at $\cos\theta = -1$, and $\mathcal{D}_1 = 4.3$ dB.
- Hypercardioid: $a_{1,1} = \frac{2}{3}$, null at $\cos\theta = -1/2$, and $\mathcal{D}_1 = 4.8$ dB.
- Supercardioid: $a_{1,1} = 2 - \sqrt{2}$, null at $\cos\theta = (1 - \sqrt{2})/(2 - \sqrt{2})$, and $\mathcal{D}_1 - 4.6$ dB.

Figure 2.2 shows these different polar patterns. What is exactly shown are the values of the magnitude squared beampattern in dB, i.e., $10\log_{10}\mathcal{B}_1^2(\theta)$.

Second-order beampatterns are described by the equation:

$$\mathcal{B}_2(\theta) = (1 - a_{2,1} - a_{2,2}) + a_{2,1}\cos\theta + a_{2,2}\cos^2\theta. \quad (2.15)$$

The second-order dipole has a null at $\cos\theta = 0$ and a one (maximum) at $\cos\theta = -1$. Replacing these values in (2.15), we find that $a_{2,1} = 0$ and $a_{2,2} = 1$. By analogy with the first-order and second-order dipoles, we define the Nth-order dipole as

$$\mathcal{B}_{\mathrm{D},N}(\theta) = \cos^N\theta, \quad (2.16)$$

[3] Another type of hypercardioid can be obtained by maximizing the directivity factor in the presence of a spherically isotropic noise field. There is not much difference, however, between the two patterns.

[4] Another type of supercardioid can be obtained by maximizing the front-to-back ratio in the presence of a spherically isotropic noise field. There is not much difference, however, between the two patterns.

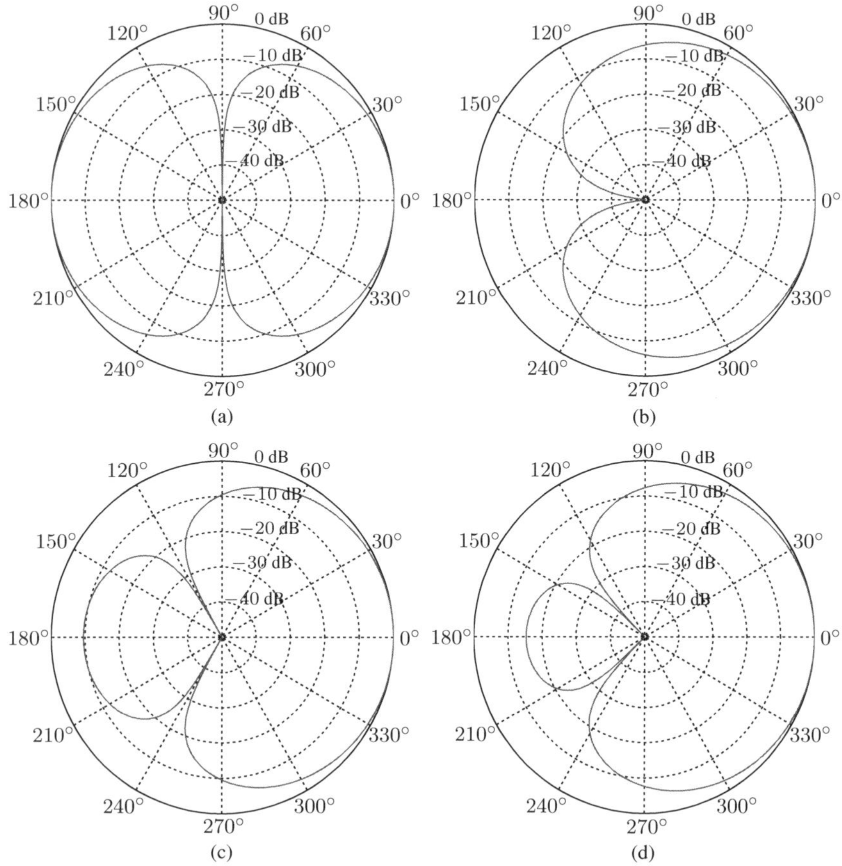

Fig. 2.2 First-order directional patterns: (a) dipole, (b) cardioid, (c) hypercardioid, and (d) supercardioid.

implying that $a_{N,N} = 1$ and $a_{N,N-1} = a_{N,N-2} = \cdots = a_{N,0} = 0$. The Nth-order dipole has only one (distinct) null (in the range $0° - 180°$) at $\theta = 90°$. The directivity indices of the second-order and third-order dipoles are, respectively, $\mathcal{D}_2 = 4.3$ dB and $\mathcal{D}_3 = 5.1$ dB.

The most well-known second-order cardioid has two nulls; one at $\cos\theta = -1$ and the other one at $\cos\theta = 0$. From these values, we easily deduce from (2.15) that $a_{2,1} = a_{2,2} = \frac{1}{2}$. By analogy with the first-order and second-order cardioids, we define the Nth-order cardioid as

$$\mathcal{B}_{\mathrm{C},N}(\theta) = \left(\frac{1}{2} + \frac{1}{2}\cos\theta\right)\cos^{N-1}\theta, \tag{2.17}$$

implying that $a_{N,N} = a_{N,N-1} = \frac{1}{2}$ and $a_{N,N-2} = a_{N,N-3} = \cdots = a_{N,0} = 0$. This Nth-order cardioid has only two distinct nulls (in the range $0° - 180°$):

one at $\theta = 90°$ and the other one at $\theta = 180°$. The directivity indices of the second-order and third-order cardioids are, respectively, $\mathcal{D}_2 = 6.6$ dB and $\mathcal{D}_3 = 7.6$ dB.

The Nth-order hypercardioid and supercardioid are characterized by the fact that they have N distinct nulls in the interval $0° < \theta < 180°$. Hence, their general beampattern is

$$\mathcal{B}_{\mathrm{HS},N}(\theta) = \prod_{n=1}^{N} \left[\varsigma_{N,n} + (1 - \varsigma_{N,n}) \cos\theta\right]. \tag{2.18}$$

Third-order beampatterns have the form

$$\begin{aligned}\mathcal{B}_3(\theta) = &(1 - a_{3,1} - a_{3,2} - a_{3,3}) + a_{3,1}\cos\theta \\ &+ a_{3,2}\cos^2\theta + a_{3,3}\cos^3\theta.\end{aligned} \tag{2.19}$$

We give the values of $a_{N,n}$ and $\mathcal{D}_N$ for some examples of hypercardioid and supercardioid [4], [6]:

- second-order hypercardioid, $a_{2,1} = \frac{2}{5}$, $a_{2,2} = \frac{4}{5}$, $\mathcal{D}_2 = 7$ dB;
- second-order supercardioid, $a_{2,1} \approx 0.484$, $a_{2,2} \approx 0.413$, $\mathcal{D}_2 = 6.3$ dB;
- third-order hypercardioid, $a_{3,1} = -\frac{4}{7}$, $a_{3,2} = \frac{4}{7}$, $a_{3,3} = \frac{8}{7}$, $\mathcal{D}_3 = 8.4$ dB; and
- third-order supercardioid, $a_{3,1} \approx 0.217$, $a_{3,2} \approx 0.475$, $a_{3,3} \approx 0.286$, $\mathcal{D}_3 = 7.2$ dB.

Figures 2.3 and 2.4 depict the different second-order and third-order directional patterns discussed above.

We are now going to show how the general definition of the beampattern given in (2.4) is very much related to the particular definition of the Nth-order directional pattern given in (2.5) for the steering vector defined in (2.1). As a consequence, the dimension (equal to the number microphones) of the vector $\mathbf{d}(\omega, \cos\theta)$ is related to the order N.

Given a function $f(x)$ such that

$$f^{(n)}(x) = \frac{d^n f(x)}{dx^n} \tag{2.20}$$

exists, the MacLaurin's series of $f(x)$ is

$$f(x) = \sum_{n=0}^{N} \frac{1}{n!} f^{(n)}(0) x^n + R_{N+1}(x), \tag{2.21}$$

where $R_{N+1}(x)$ is some remainder with

$$\lim_{N\to\infty} R_N(x) = 0. \tag{2.22}$$

We deduce that the MacLaurin's series for the exponential is

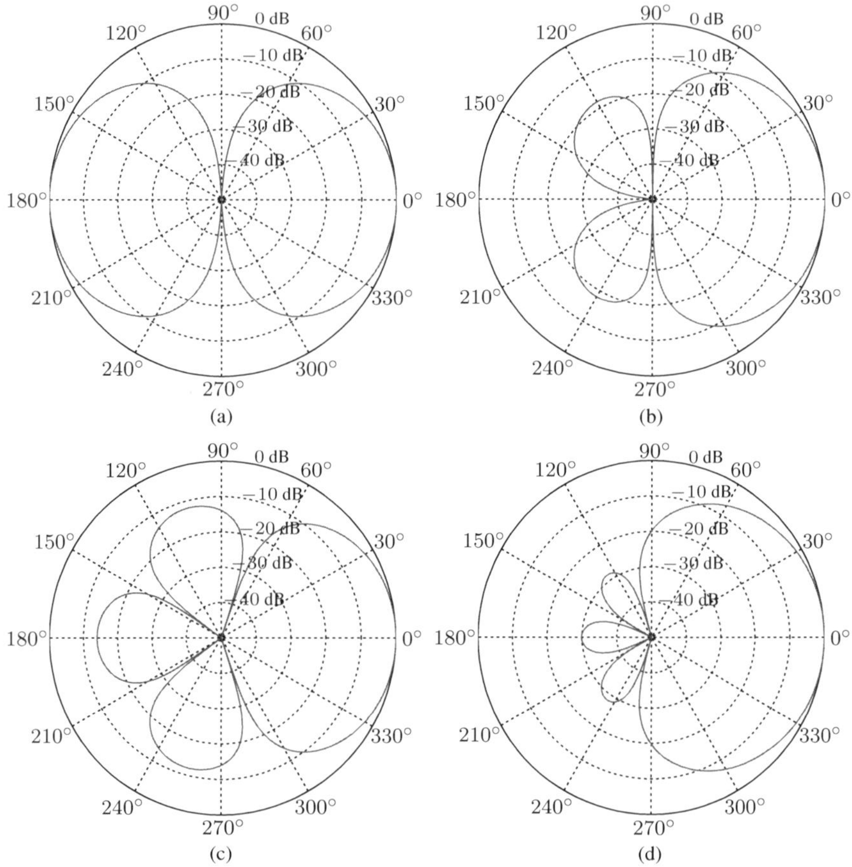

Fig. 2.3 Second-order directional patterns: (a) dipole, (b) cardioid, (c) hypercardioid, and (d) supercardioid.

$$e^x = \sum_{n=0}^{N} \frac{1}{n!} x^n + R_{N+1}(x). \tag{2.23}$$

Substituting $x = \jmath(m-1)\omega\tau_0 \cos\theta$ in (2.23) and neglecting the remainder, we find that

$$e^{\jmath(m-1)\omega\tau_0 \cos\theta} \approx \sum_{n=0}^{N} \frac{1}{n!} \left[\jmath(m-1)\omega\tau_0 \cos\theta\right]^n . \tag{2.24}$$

Using (2.24) in the general definition of the beampattern, we obtain

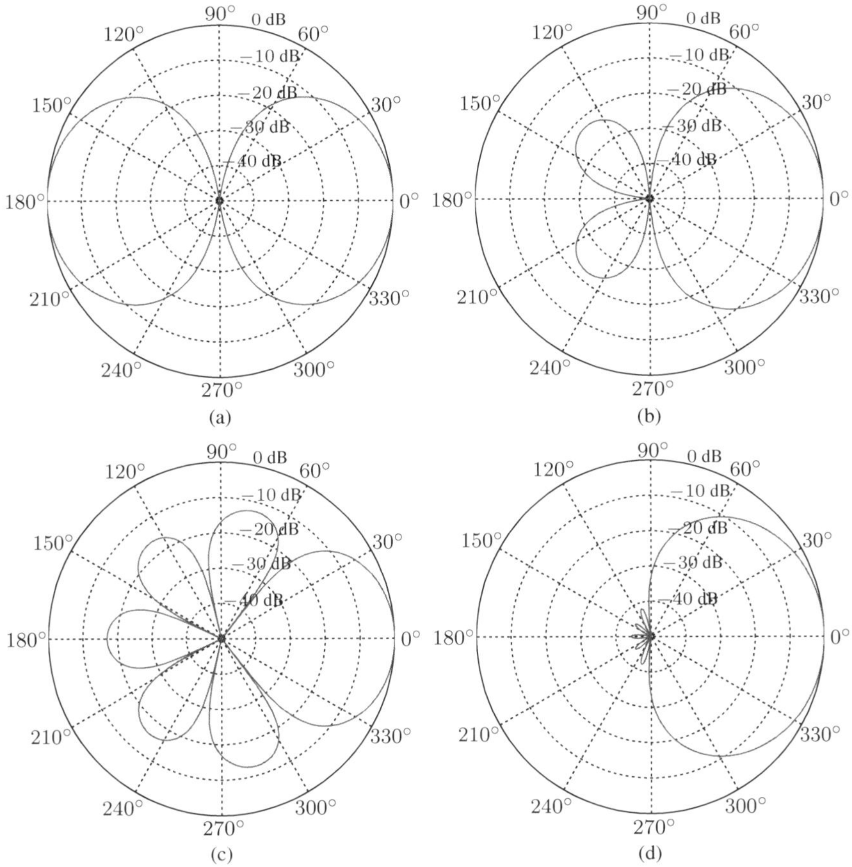

Fig. 2.4 Third-order directional patterns: (a) dipole, (b) cardioid, (c) hypercardioid, and (d) supercardioid.

$$
\begin{aligned}
\mathcal{B}\left[\mathbf{h}\left(\omega\right),\theta\right] &= \sum_{m=1}^{M} H_m\left(\omega\right) e^{\jmath(m-1)\omega\tau_0\cos\theta} \\
&\approx \sum_{m=1}^{M} H_m\left(\omega\right) \sum_{n=0}^{N} \frac{1}{n!}\left[\jmath(m-1)\omega\tau_0\cos\theta\right]^n \\
&\approx \sum_{n=0}^{N} \cos^n\theta \left[\frac{\left(\jmath\omega\tau_0\right)^n}{n!} \sum_{m=1}^{M} (m-1)^n H_m\left(\omega\right)\right] \\
&\approx \sum_{n=0}^{N} a_{N,n}\cos^n\theta = \mathcal{B}_N\left(\theta\right),
\end{aligned}
\tag{2.25}
$$

where

$$a_{N,n} \approx \frac{(\jmath\omega\tau_0)^n}{n!} \sum_{m=1}^{M} (m-1)^n H_m(\omega). \tag{2.26}$$

We observe from (2.25) that as long as $e^{\jmath(m-1)\omega\tau_0\cos\theta}$ can be approximated by a MacLaurin's series of order N (that is why the microphone spacing should be small), which includes derivatives up to the order N, we can build Nth-order differential arrays. We also observe from (2.26) that the gains $H_m(\omega)$, $m = 1, 2, \ldots, M$, can be determined given the coefficients $a_{N,n}$, $n = 0, 1, \ldots, N$. The least-squares solution ($N+1 > M$) is not appropriate since not only the beampatterns will be highly frequency dependent (it is very hard, if not impossible, to numerically approximate a derivative of order N with a smaller number of points, M) but it is also very hard to have exact nulls in some specific directions and a one at $\theta = 0°$. The minimum-norm solution ($N+1 < M$) is a good choice from both theoretical and practical viewpoints; this concept will be elaborated in Chapter 6. But for all the other chapters, it will always be assumed that the design of an Nth-order differential array requires $N+1$ microphones.

2.3 Gain in Signal-to-Noise Ratio (SNR)

The first microphone serves as the reference and we recall that the desired signal comes from the angle $\theta = 0°$. In this case, the mth microphone signal is given by

$$Y_m(\omega) = e^{-\jmath(m-1)\omega\tau_0} X(\omega) + V_m(\omega), \; m = 1, 2, \ldots, M, \tag{2.27}$$

where $X(\omega)$ is the desired signal and $V_m(\omega)$ is the additive noise at the mth microphone. In a vector form, (2.27) becomes

$$\begin{aligned} \mathbf{y}(\omega) &= \left[Y_1(\omega) \; Y_2(\omega) \cdots Y_M(\omega) \right]^T \\ &= \mathbf{d}(\omega, \cos 0°) X(\omega) + \mathbf{v}(\omega), \end{aligned} \tag{2.28}$$

where the noise signal vector, $\mathbf{v}(\omega)$, is defined similarly to $\mathbf{y}(\omega)$.

The beamformer output is simply

$$\begin{aligned} Z(\omega) &= \sum_{m=1}^{M} H_m^*(\omega) Y_m(\omega) \\ &= \mathbf{h}^H(\omega) \mathbf{y}(\omega) \\ &= \mathbf{h}^H(\omega) \mathbf{d}(\omega, \cos 0°) X(\omega) + \mathbf{h}^H(\omega) \mathbf{v}(\omega), \end{aligned} \tag{2.29}$$

where $Z(\omega)$ is supposed to be the estimate of the desired signal, $X(\omega)$.

We define the input signal-to-noise ratio (SNR) as

$$\text{iSNR}(\omega) = \frac{\phi_X(\omega)}{\phi_{V_1}(\omega)}, \tag{2.30}$$

where $\phi_X(\omega) = E\left[|X(\omega)|^2\right]$ and $\phi_{V_1}(\omega) = E\left[|V_1(\omega)|^2\right]$ are the variances of $X(\omega)$ and $V_1(\omega)$, respectively.

The output SNR is defined as

$$\begin{aligned} \text{oSNR}[\mathbf{h}(\omega)] &= \phi_X(\omega)\frac{\left|\mathbf{h}^H(\omega)\mathbf{d}(\omega,\cos 0^\circ)\right|^2}{\mathbf{h}^H(\omega)\mathbf{\Phi}_\mathbf{v}(\omega)\mathbf{h}(\omega)} \\ &= \frac{\phi_X(\omega)}{\phi_{V_1}(\omega)} \cdot \frac{\left|\mathbf{h}^H(\omega)\mathbf{d}(\omega,\cos 0^\circ)\right|^2}{\mathbf{h}^H(\omega)\mathbf{\Gamma}_\mathbf{v}(\omega)\mathbf{h}(\omega)}, \end{aligned} \tag{2.31}$$

where

$$\mathbf{\Phi}_\mathbf{v}(\omega) = E\left[\mathbf{v}(\omega)\mathbf{v}^H(\omega)\right] \tag{2.32}$$

and

$$\mathbf{\Gamma}_\mathbf{v}(\omega) = \frac{\mathbf{\Phi}_\mathbf{v}(\omega)}{\phi_{V_1}(\omega)} \tag{2.33}$$

are the correlation and pseudo-coherence matrices of $\mathbf{v}(\omega)$, respectively.

The definition of the gain in SNR is easily derived from the two previous definitions, i.e.,

$$\begin{aligned} \mathcal{G}[\mathbf{h}(\omega)] &= \frac{\text{oSNR}[\mathbf{h}(\omega)]}{\text{iSNR}(\omega)} \\ &= \frac{\left|\mathbf{h}^H(\omega)\mathbf{d}(\omega,\cos 0^\circ)\right|^2}{\mathbf{h}^H(\omega)\mathbf{\Gamma}_\mathbf{v}(\omega)\mathbf{h}(\omega)}. \end{aligned} \tag{2.34}$$

Assume that the matrix $\mathbf{\Gamma}_\mathbf{v}(\omega)$ is nonsingular. In this case, for any two vectors $\mathbf{h}(\omega)$ and $\mathbf{d}(\omega,\cos 0^\circ)$, we have

$$\begin{aligned} \left|\mathbf{h}^H(\omega)\mathbf{d}(\omega,\cos 0^\circ)\right|^2 \leq &\left[\mathbf{h}^H(\omega)\mathbf{\Gamma}_\mathbf{v}(\omega)\mathbf{h}(\omega)\right] \times \\ &\left[\mathbf{d}^H(\omega,\cos 0^\circ)\mathbf{\Gamma}_\mathbf{v}^{-1}(\omega)\mathbf{d}(\omega,\cos 0^\circ)\right], \end{aligned} \tag{2.35}$$

with equality if and only if $\mathbf{h}(\omega) \propto \mathbf{\Gamma}_\mathbf{v}^{-1}(\omega)\mathbf{d}(\omega,\cos 0^\circ)$. Using the inequality (2.35) in (2.34), we deduce an upper bound for the gain:

$$\begin{aligned} \mathcal{G}[\mathbf{h}(\omega)] &\leq \mathbf{d}^H(\omega,\cos 0^\circ)\mathbf{\Gamma}_\mathbf{v}^{-1}(\omega)\mathbf{d}(\omega,\cos 0^\circ) \\ &\leq \text{tr}\left[\mathbf{\Gamma}_\mathbf{v}^{-1}(\omega)\right]\text{tr}\left[\mathbf{d}(\omega,\cos 0^\circ)\mathbf{d}^H(\omega,\cos 0^\circ)\right] \\ &\leq M\text{tr}\left[\mathbf{\Gamma}_\mathbf{v}^{-1}(\omega)\right], \end{aligned} \tag{2.36}$$

where tr[·] is the trace of a square matrix. We observe how the gain is upper bounded [as long as $\mathbf{\Gamma_v}(\omega)$ is nonsingular] and depends on the number of microphones as well as on the nature of the noise.

In our context, the distortionless constraint is desired, i.e.,

$$\mathbf{h}^H(\omega)\,\mathbf{d}(\omega, \cos 0^\circ) = 1. \tag{2.37}$$

As a consequence, it is easy to see that the filter:

$$\mathbf{h}_{\max}(\omega) = \frac{\mathbf{\Gamma_v}^{-1}(\omega)\,\mathbf{d}(\omega, \cos 0^\circ)}{\mathbf{d}^H(\omega, \cos 0^\circ)\,\mathbf{\Gamma_v}^{-1}(\omega)\,\mathbf{d}(\omega, \cos 0^\circ)} \tag{2.38}$$

maximizes the gain, which is given by

$$\mathcal{G}_{\max}(\omega) = \mathbf{d}^H(\omega, \cos 0^\circ)\,\mathbf{\Gamma_v}^{-1}(\omega)\,\mathbf{d}(\omega, \cos 0^\circ). \tag{2.39}$$

We are interested in three types of noise.

- The temporally and spatially white noise with the same variance at all microphones[5]. In this case, $\mathbf{\Gamma_v}(\omega) = \mathbf{I}_M$, where $\mathbf{I}_M$ is the $M \times M$ identity matrix. Therefore, the white noise gain is

$$\begin{aligned}\mathcal{G}_{\text{wn}}[\mathbf{h}(\omega)] &= \frac{\left|\mathbf{h}^H(\omega)\,\mathbf{d}(\omega, \cos 0^\circ)\right|^2}{\mathbf{h}^H(\omega)\,\mathbf{h}(\omega)} \\ &= \frac{1}{\mathbf{h}^H(\omega)\,\mathbf{h}(\omega)},\end{aligned} \tag{2.40}$$

  where in the second line of (2.40), the distortionless constraint is assumed. For

$$\mathbf{h}(\omega) = \frac{\mathbf{d}(\omega, \cos 0^\circ)}{M}, \tag{2.41}$$

  we find the maximum possible gain, which is

$$\mathcal{G}_{\text{wn},\max}(\omega) = M. \tag{2.42}$$

  In general, the white noise gain of an Nth-order DMA is

$$\mathcal{G}_{\text{wn},N}[\mathbf{h}(\omega)] = \frac{1}{\mathbf{h}^H(\omega)\,\mathbf{h}(\omega)} \leq M. \tag{2.43}$$

  We will see how the white noise may be amplified by DMAs, especially at low frequencies.
- The diffuse noise[6], where

[5] This noise models well the sensor noise.

[6] This situation corresponds to the spherically isotropic noise field.

$$
\begin{aligned}
\left[\boldsymbol{\Gamma}_{\mathbf{v}}(\omega)\right]_{ij} = \left[\boldsymbol{\Gamma}_{\mathrm{dn}}(\omega)\right]_{ij} &= \frac{\sin\left[\omega(j-i)\tau_0\right]}{\omega(j-i)\tau_0} \\
&= \mathrm{sinc}\left[\omega(j-i)\tau_0\right]. \qquad (2.44)
\end{aligned}
$$

In this scenario, the gain in SNR, $\mathcal{G}_{\mathrm{dn}}\left[\mathbf{h}(\omega)\right]$, is called the directivity factor and the directivity index is simply defined as [2], [4]

$$
\mathcal{D}\left[\mathbf{h}(\omega)\right] = 10\log_{10}\mathcal{G}_{\mathrm{dn}}\left[\mathbf{h}(\omega)\right]. \qquad (2.45)
$$

With diffuse noise, the filter $\mathbf{h}(\omega)$ is often found by maximizing the directivity factor. As a result, the optimal filter is given by (2.38).

- The noise comes from a point source at the angle θ_{n}. In this case, the pseudo-coherence matrix is

$$
\boldsymbol{\Gamma}_{\mathbf{v}}(\omega) = \mathbf{d}(\omega, \cos\theta_{\mathrm{n}})\,\mathbf{d}^H(\omega, \cos\theta_{\mathrm{n}}), \qquad (2.46)
$$

where

$$
\mathbf{d}(\omega, \cos\theta_{\mathrm{n}}) = \left[1 \;\; e^{-\jmath\omega\tau_0\cos\theta_{\mathrm{n}}} \;\; \cdots \;\; e^{-\jmath(M-1)\omega\tau_0\cos\theta_{\mathrm{n}}}\right]^T \qquad (2.47)
$$

is the steering vector of the noise source. We observe from (2.46) that the pseudo-coherence matrix is singular. In fact, this is the only possibility where the gain in SNR, $\mathcal{G}_{\mathrm{ns}}\left[\mathbf{h}(\omega)\right]$, is not upper bounded and can go to infinity. We deduce that this gain is

$$
\begin{aligned}
\mathcal{G}_{\mathrm{ns}}\left[\mathbf{h}(\omega)\right] &= \frac{\left|\mathbf{h}^H(\omega)\,\mathbf{d}(\omega, \cos 0^\circ)\right|^2}{\left|\mathbf{h}^H(\omega)\,\mathbf{d}(\omega, \cos\theta_{\mathrm{n}})\right|^2} \\
&= \frac{1}{\left|\mathbf{h}^H(\omega)\,\mathbf{d}(\omega, \cos\theta_{\mathrm{n}})\right|^2}. \qquad (2.48)
\end{aligned}
$$

When the noise and desired signals come from the same direction, i.e., when $\theta_{\mathrm{n}} = 0^\circ$, then there is no possible gain, i.e., $\mathcal{G}_{\mathrm{ns}}\left[\mathbf{h}(\omega)\right] = 1,\ \forall\mathbf{h}(\omega)$. We also deduce the gain of an Nth-order DMA:

$$
\mathcal{G}_{\mathrm{ns},N}(\theta_{\mathrm{n}}) = \frac{1}{\left|\mathcal{B}_N(\theta_{\mathrm{n}})\right|^2}. \qquad (2.49)
$$

Figures 2.5, 2.6, and 2.7 depict this gain, as a function of the direction of the noise, for the different first-order, second-order, and third-order patterns (dipole, cardioid, hypercardioid, and supercardioid).

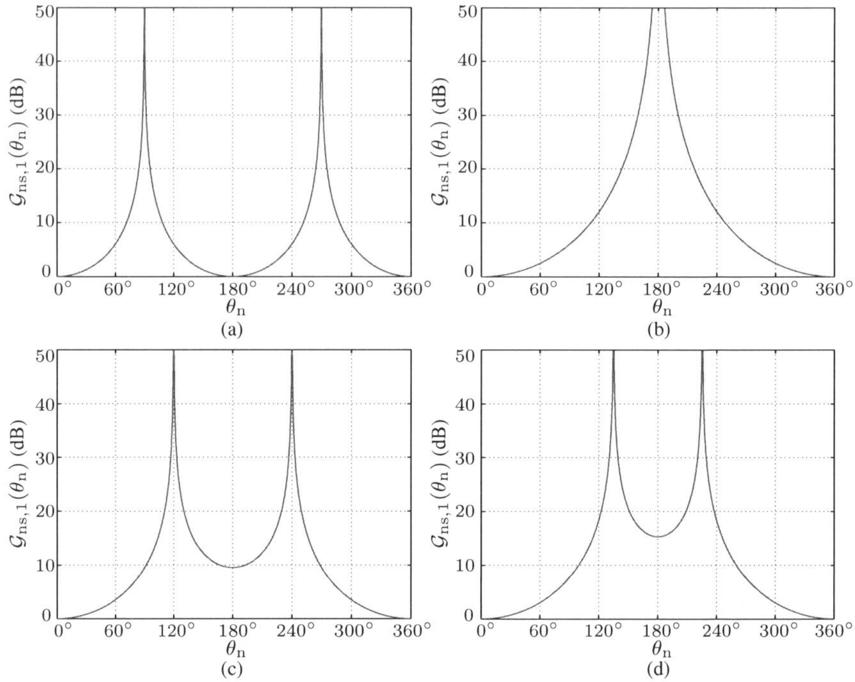

Fig. 2.5 Gain in SNR as a function of the direction (θ_n) of the point noise source for the first-order DMA: (a) dipole, (b) cardioid, (c) hypercardioid, and (d) supercardioid.

2.4 Vandermonde Matrix

Given the definition of the steering vector and combining steering vectors for different angles in a matrix, we obtain the Vandermonde structure. Therefore, it is extremely useful to exploit the structure of this matrix.

A Vandermonde matrix of size $M \times M$ has the form:

$$\mathbf{V}_M = \begin{bmatrix} 1 & v_1 & v_1^2 & \cdots & v_1^{M-1} \\ 1 & v_2 & v_2^2 & \cdots & v_2^{M-1} \\ 1 & v_3 & v_3^2 & \cdots & v_3^{M-1} \\ \vdots & \vdots & \vdots & \ddots & \vdots \\ 1 & v_M & v_M^2 & \cdots & v_M^{M-1} \end{bmatrix}. \tag{2.50}$$

It can be shown that the determinant of $\mathbf{V}_M$ is

$$\det(\mathbf{V}_M) = \prod_{j>i} (v_j - v_i). \tag{2.51}$$

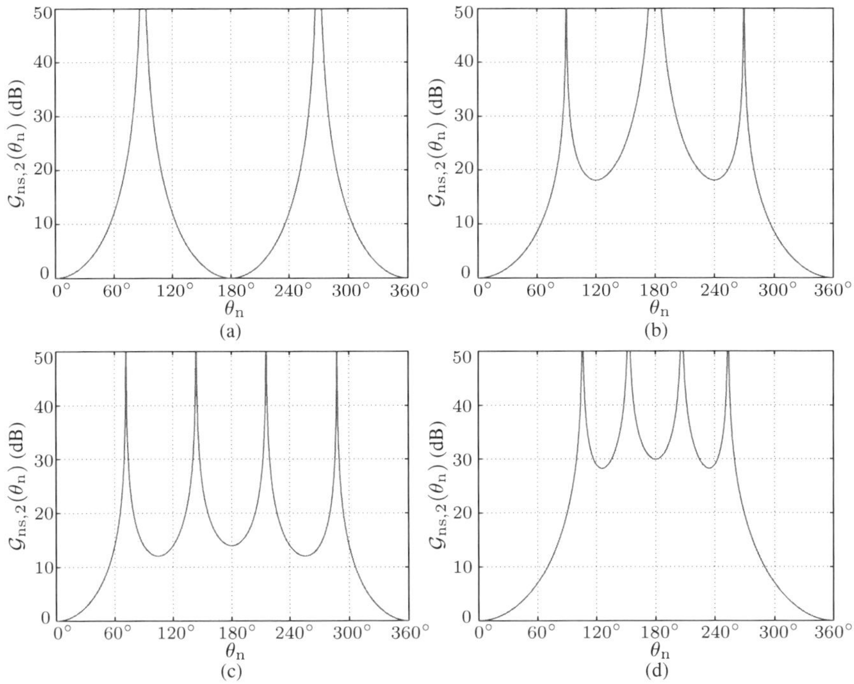

Fig. 2.6 Gain in SNR as a function of the direction (θ_{n}) of the point noise source for the second-order DMA: (a) dipole, (b) cardioid, (c) hypercardioid, and (d) supercardioid.

As a consequence, as long as the values of v_m are all distinct, the matrix $\mathbf{V}_M$ is nonsingular.

It will be important to have a closed-form expression of the inverse of the Vandermonde matrix. For that, we will use the decomposition proposed in [8]:

$$\mathbf{V}_M^{-1} = \mathbf{U}_M \mathbf{L}_M, \tag{2.52}$$

where $\mathbf{U}_M$ and $\mathbf{L}_M$ are upper and lower triangular matrices, respectively.

The elements l_{ij} of $\mathbf{L}_M$ are given by the relations:

$$l_{ij} = \begin{cases} 0, & i < j \\ 1, & i = j = 1 \\ \prod_{p=1, p \neq j}^{i} \dfrac{1}{v_j - v_p}, & \text{otherwise} \end{cases}. \tag{2.53}$$

It is proved in [8] that the elements u_{ij} of $\mathbf{U}_M$ are given by the definition:

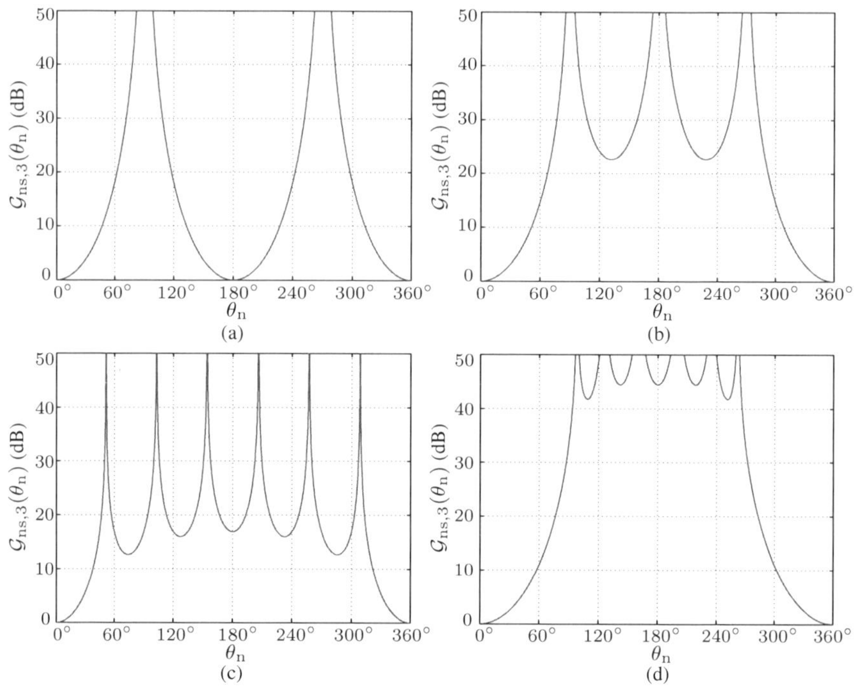

Fig. 2.7 Gain in SNR as a function of the direction ($\theta_{\rm n}$) of the point noise source for the third-order DMA: (a) dipole, (b) cardioid, (c) hypercardioid, and (d) supercardioid.

$$u_{ij} = \begin{cases} 1, & j = i \\ 0, & j = 1 \\ u_{i-1,j-1} - u_{i,j-1}v_{j-1}, & \text{otherwise} \end{cases}, \tag{2.54}$$

where

$$u_{0j} = 0. \tag{2.55}$$

References

1. G. W. Elko, J. E. West, and S. Thompson, "Differential and gradient microphone arrays," *J. Acoust. Soc. Am.*, vol. 114, pp. 2426–2426, Oct. 2003.
2. G. W. Elko and J. Meyer, "Microphone arrays," in *Springer Handbook of Speech Processing*, J. Benesty, M. M. Sondhi, and Y. Huang, Eds. Berlin: Germany, Springer-Verlag, 2008, Chapter 50, Part I, pp. 1021–1041.
3. M. Buck, "Aspects of first-order differential microphone arrays in the presence of sensor imperfections," *European Trans. Telecommunications*, vol. 13, pp. 115-122, Mar.-Apr. 2002.
4. G. W. Elko, "Superdirectional microphone arrays," in *Acoustic Signal Processing for Telecommunication*, S. L. Gay and J. Benesty, Eds. Boston, MA: Kluwer Academic

Publishers, 2000, Chapter 10, pp. 181–237.
5. M. Buck and M. Rößler, "First order differential microphone arrays for automative applications," in *Proc. IWAENC*, 2001.
6. E. De Sena, H. Hacihabiboğlu, and Z. Cvetković, "On the design and implementation of higher-order differential microphones," *IEEE Trans. Audio, Speech, Language Process.*, vol. 20, pp. 162–174, Jan. 2012.
7. R. N. Marshall and W. R. Harry, "A new microphone providing uniform directivity over an extended frequency range," *J. Acoust. Soc. Am.*, vol. 12, pp. 481–497, 1941.
8. L. R. Turner, "Inverse of the Vandermonde matrix with applications," *NASA TN D-3547*, Aug. 1966.

Chapter 3
Study and Design of First-Order Differential Arrays

This chapter is dedicated to the study and design of first-order differential microphone arrays. We explore the dipole, cardioid, subcardioid, hypercardioid, and supercardioid.

3.1 Principle

In this work, the proposed approach to design DMAs is based on the obvious observation that any ideal frequency-independent directional pattern has a one at the angle $\theta = 0°$ and a number of nulls in some specific directions. First-order DMAs are designed with two microphones. In this case, we have exactly two constraints to fulfill. The first constraint is the distortionless response (a one at the angle $\theta = 0°$) and the second constraint is a null in the interval $0° < \theta \leq 180°$. Thus, these two constraints can be written as

$$\mathbf{d}^H(\omega, \cos 0°)\,\mathbf{h}(\omega) = \mathbf{d}^H(\omega, 1)\,\mathbf{h}(\omega) = 1, \tag{3.1}$$

$$\mathbf{d}^H(\omega, \alpha_{1,1})\,\mathbf{h}(\omega) = 0, \tag{3.2}$$

where $\alpha_{1,1} = \cos\theta_{1,1}$ is given by design (a null at the angle $\theta_{1,1}$) with $-1 \leq \alpha_{1,1} < 1$. We can express (3.1)–(3.2) as

$$\begin{bmatrix} \mathbf{d}^H(\omega, 1) \\ \mathbf{d}^H(\omega, \alpha_{1,1}) \end{bmatrix} \mathbf{h}(\omega) = \begin{bmatrix} 1 \\ 0 \end{bmatrix}. \tag{3.3}$$

The matrix involved in (3.3) is recognized to be a 2×2 Vandermonde matrix. Therefore, to design any first-order DMA, we only need to solve (3.3) and the obtained solution is the best we can get from a mathematical point of view.

J. Benesty and J. Chen, *Study and Design of Differential Microphone Arrays*, Springer Topics in Signal Processing 6, DOI: 10.1007/978-3-642-33753-6_3,

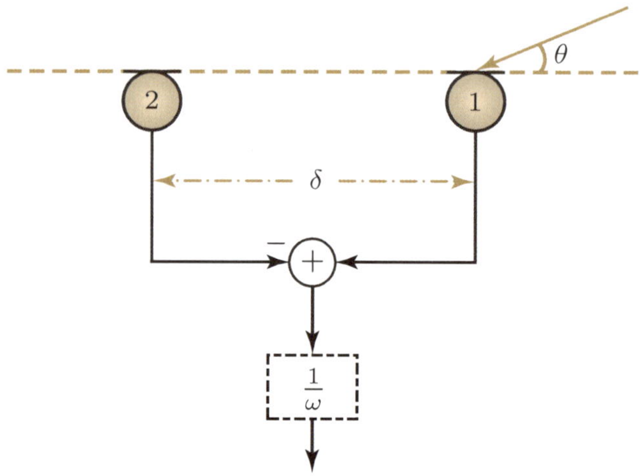

Fig. 3.1 Implementation of the first-order dipole.

3.2 First-Order Dipole

The first-order dipole has a one at the angle 0° and a null at the angle 90° (i.e., $\alpha_{1,1} = 0$). Hence, the system of two linear equations (3.3) becomes

$$\begin{bmatrix} 1 & e^{\jmath\omega\tau_0} \\ 1 & 1 \end{bmatrix} \mathbf{h}(\omega) = \begin{bmatrix} 1 \\ 0 \end{bmatrix}, \tag{3.4}$$

for which the solution is

$$\mathbf{h}(\omega) = \frac{1}{1 - e^{\jmath\omega\tau_0}} \begin{bmatrix} 1 \\ -1 \end{bmatrix}. \tag{3.5}$$

We recall that $\tau_0 = \delta/c$. Assuming that the sensor spacing is much smaller than the acoustic wavelength, we can use the approximation:

$$e^x \approx 1 + x \tag{3.6}$$

in (3.5) to get

$$\mathbf{h}(\omega) \approx \jmath \frac{1}{\omega\tau_0} \begin{bmatrix} 1 \\ -1 \end{bmatrix}. \tag{3.7}$$

The frequency-independent constant $\mathcal{C} = \jmath \frac{1}{\tau_0}$ has no importance; therefore, $\mathbf{h}(\omega)$ is simplified to the equivalent filter:

$$\mathbf{h}'(\omega) = \frac{1}{\omega} \begin{bmatrix} 1 \\ -1 \end{bmatrix}, \tag{3.8}$$

where the term $1/\omega$ in (3.8) is simply a first-order low-pass filter. Figure 3.1 illustrates the implementation of the first-order dipole.

Now, the beampattern is

$$\begin{aligned}\mathcal{B}\left[\mathbf{h}'(\omega), \theta\right] &= \mathbf{d}^H(\omega, \cos\theta)\, \mathbf{h}'(\omega) \\ &= \frac{1}{\omega}\left(1 - e^{\jmath\omega\tau_0\cos\theta}\right). \end{aligned} \tag{3.9}$$

Figure 3.2 displays the patterns from (3.9) for several frequencies and two values of δ. Applying the approximation (3.6) to (3.9) leads to

$$\mathcal{B}\left[\mathbf{h}'(\omega), \theta\right] \approx \frac{1}{\mathcal{C}} \cos\theta. \tag{3.10}$$

Obviously, we recognize the pattern of the first-order dipole since $1/\mathcal{C}$ is simply a constant.

The white noise gain is

$$\begin{aligned}\mathcal{G}_{\mathrm{wn},1}\left[\mathbf{h}'(\omega)\right] &= \frac{\left|\mathbf{h}'^H(\omega)\,\mathbf{d}(\omega, 1)\right|^2}{\mathbf{h}'^H(\omega)\,\mathbf{h}'(\omega)} \\ &= \frac{1}{2}\left|1 - e^{\jmath\omega\tau_0}\right|^2 \\ &= 1 - \cos(\omega\tau_0). \end{aligned} \tag{3.11}$$

Figure 3.3 gives plots of $\mathcal{G}_{\mathrm{wn},1}\left[\mathbf{h}'(\omega)\right]$ from (3.11), as a function of frequency, for different values of δ.

The white noise is amplified if

$$\mathcal{G}_{\mathrm{wn},1}\left[\mathbf{h}'(\omega)\right] < 1. \tag{3.12}$$

We observe from (3.11) that the sensor spacing must be chosen large, especially for low frequencies, if we don't want to amplify the white (or sensor) noise. But a large value of δ is in contradiction with the DMA assumption, which states that δ should be small. Therefore, there is always a tradeoff between white noise amplification (especially at low frequencies) and frequency-independent directional pattern at high frequencies. Hence, the sensor spacing should be selected according to this compromise. To better illustrate this aspect, let us approximate (3.11) for small values of $\omega\tau_0$. We get

$$\mathcal{G}_{\mathrm{wn},1}\left[\mathbf{h}'(\omega)\right] \approx \frac{(\omega\tau_0)^2}{2}. \tag{3.13}$$

We see that the gain is greater than 1 if $\omega\tau_0 > \sqrt{2}$, which is somewhat in contradiction with the condition given in (2.2).

The directivity factor (or the gain in SNR with diffuse noise) is

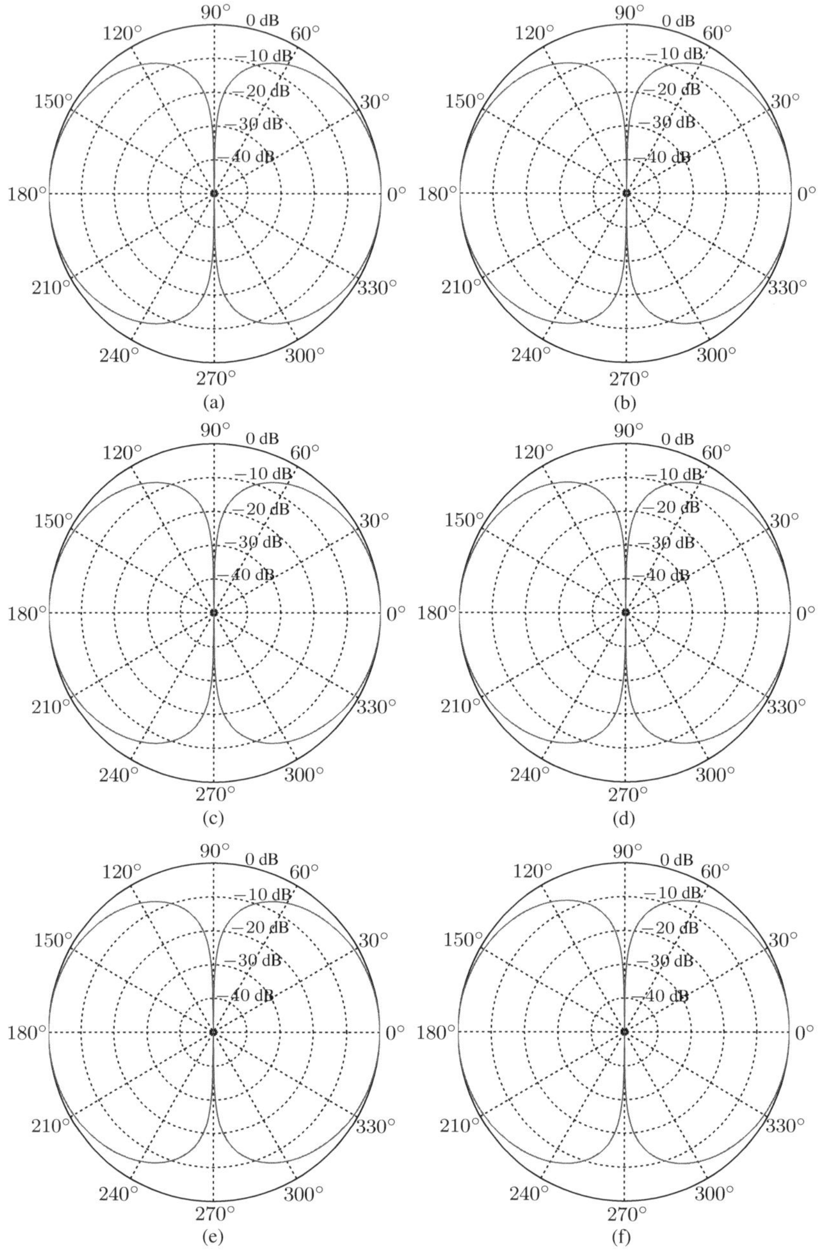

Fig. 3.2 Patterns of the first-order dipole for several frequencies and two values of δ: (a) $f = 0.5$ kHz, $\delta = 1$ cm, (b) $f = 0.5$ kHz, $\delta = 2$ cm, (c) $f = 1$ kHz, $\delta = 1$ cm, (d) $f = 1$ kHz, $\delta = 2$ cm, (e) $f = 3$ kHz, $\delta = 1$ cm, and (f) $f = 3$ kHz, $\delta = 2$ cm.

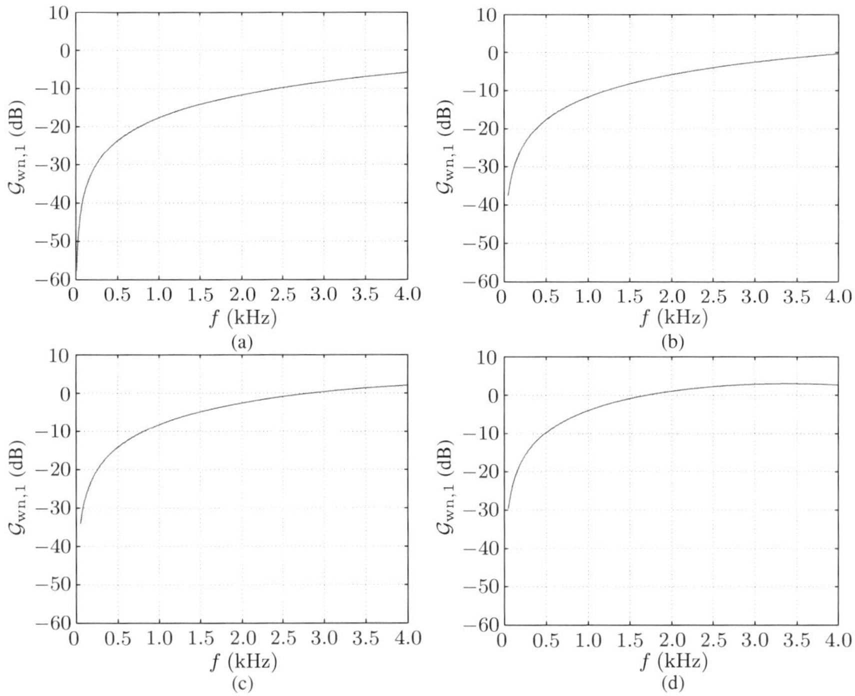

Fig. 3.3 The white noise gain of the first-order dipole, as a function of frequency, for different values of δ: (a) $\delta = 1$ cm, (b) $\delta = 2$ cm, (c) $\delta = 3$ cm, and (d) $\delta = 5$ cm.

$$\begin{aligned}\mathcal{G}_{\mathrm{dn},1}\left[\mathbf{h}'(\omega)\right] &= \frac{\left|\mathbf{h}'^{H}(\omega)\,\mathbf{d}(\omega,1)\right|^{2}}{\mathbf{h}'^{H}(\omega)\,\boldsymbol{\Gamma}_{\mathrm{dn}}(\omega)\,\mathbf{h}'(\omega)} \\ &= \frac{\left|1 - e^{\jmath\omega\tau_0}\right|^2}{2\left[1 - \mathrm{sinc}\left(\omega\tau_0\right)\right]} \\ &= \frac{1 - \cos\left(\omega\tau_0\right)}{1 - \mathrm{sinc}\left(\omega\tau_0\right)}. \end{aligned} \tag{3.14}$$

Figure 3.4 gives plots of $\mathcal{G}_{\mathrm{dn},1}\left[\mathbf{h}'(\omega)\right]$ from (3.14), as a function of frequency, for different values of δ.

We easily see that

$$\mathcal{G}_{\mathrm{dn},1}\left[\mathbf{h}'(\omega)\right] > \mathcal{G}_{\mathrm{wn},1}\left[\mathbf{h}'(\omega)\right]. \tag{3.15}$$

Also, for small values of $\omega\tau_0$, (3.14) becomes

$$\begin{aligned}\mathcal{G}_{\mathrm{dn},1}\left[\mathbf{h}'(\omega)\right] &\approx \frac{\left(\omega\tau_0\right)^2}{2} \cdot \frac{3!}{\left(\omega\tau_0\right)^2} \\ &\approx 3, \end{aligned} \tag{3.16}$$

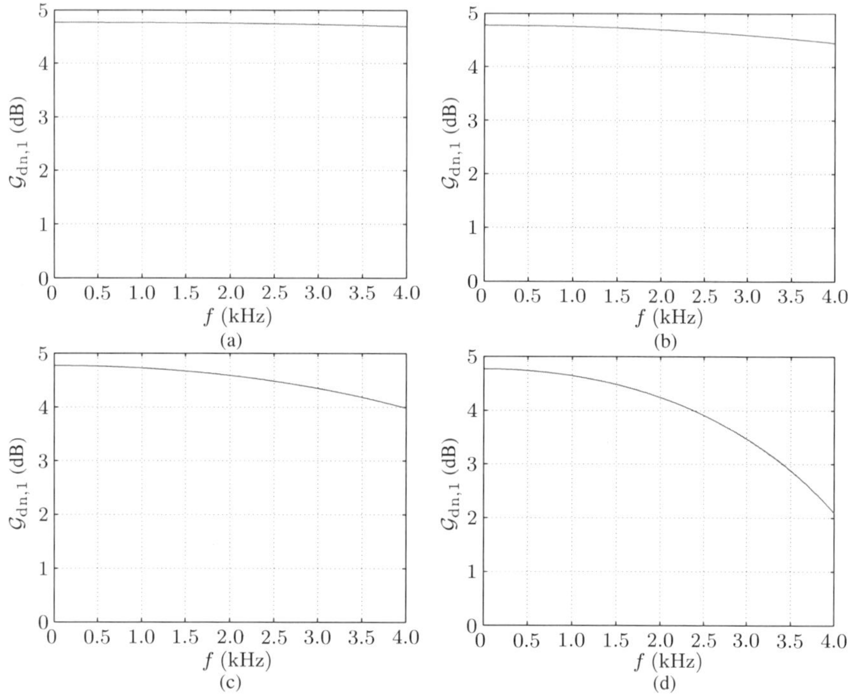

Fig. 3.4 The directivity factor of the first-order dipole, as a function of frequency, for different values of δ: (a) $\delta = 1$ cm, (b) $\delta = 2$ cm, (c) $\delta = 3$ cm, and (d) $\delta = 5$ cm.

which corresponds exactly, as expected, to the theoretical value of the directivity factor for the dipole with diffuse noise [1].

For a point noise source, the gain is

$$\mathcal{G}_{\mathrm{ns},1}\left[\mathbf{h}'(\omega)\right] = \frac{\left|\mathbf{h}'^{H}(\omega)\,\mathbf{d}(\omega,1)\right|^2}{\left|\mathbf{h}'^{H}(\omega)\,\mathbf{d}(\omega,\cos\theta_{\mathrm{n}})\right|^2} = \frac{\left|1 - e^{\jmath\omega\tau_0}\right|^2}{\left|1 - e^{\jmath\omega\tau_0\cos\theta_{\mathrm{n}}}\right|^2} = \frac{1-\cos(\omega\tau_0)}{1-\cos(\omega\tau_0\cos\theta_{\mathrm{n}})}. \tag{3.17}$$

Therefore, for $\theta_{\mathrm{n}} = 0°$, $\mathcal{G}_{\mathrm{ns},1}\left[\mathbf{h}'(\omega)\right] = 1$, $\forall f$ and for $\theta_{\mathrm{n}} = 90°$, $\mathcal{G}_{\mathrm{ns},1}\left[\mathbf{h}'(\omega)\right] = \infty$, $\forall f$. Figure 3.5 shows $\mathcal{G}_{\mathrm{ns},1}\left[\mathbf{h}'(\omega)\right]$, as a function of θ_{n}, for several frequencies and two values of δ. With the conventional approximation, we find that

$$\mathcal{G}_{\mathrm{ns},1}\left[\mathbf{h}'(\omega)\right] \approx \frac{1}{\cos^2\theta_{\mathrm{n}}}, \tag{3.18}$$

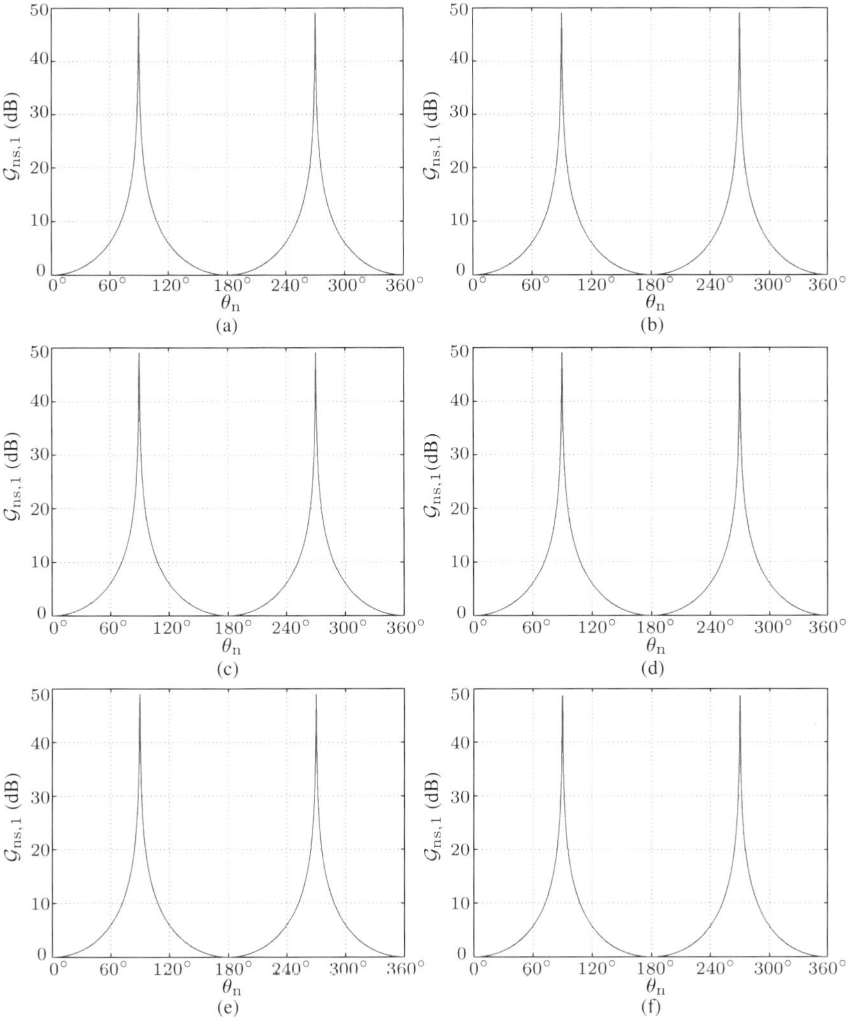

Fig. 3.5 The gain of the first-order dipole for a point noise source, as a function of θ_{n}, for several frequencies and two values of δ: (a) $f = 0.5$ kHz, $\delta = 1$ cm, (b) $f = 0.5$ kHz, $\delta = 2$ cm, (c) $f = 1$ kHz, $\delta = 1$ cm, (d) $f = 1$ kHz, $\delta = 2$ cm, (e) $f = 3$ kHz, $\delta = 1$ cm, and (f) $f = 3$ kHz, $\delta = 2$ cm.

which corresponds to the theoretical gain of the first-order dipole.

3.3 First-Order Cardioid

In the first-order cardioid, there is a one at the angle 0° and a null at the angle 180° (i.e., $\alpha_{1,1} = -1$). Therefore, our linear system of two equations is

$$\begin{bmatrix} 1 & e^{\jmath\omega\tau_0} \\ 1 & e^{-\jmath\omega\tau_0} \end{bmatrix} \mathbf{h}(\omega) = \begin{bmatrix} 1 \\ 0 \end{bmatrix}. \tag{3.19}$$

We easily find that the solution is

$$\mathbf{h}(\omega) = \frac{1}{1 - e^{\jmath 2\omega\tau_0}} \begin{bmatrix} 1 \\ -e^{\jmath\omega\tau_0} \end{bmatrix}. \tag{3.20}$$

Using (3.6) to approximate $1 - e^{\jmath 2\omega\tau_0}$, we obtain

$$\mathbf{h}(\omega) \approx \jmath \frac{1}{2\omega\tau_0} \begin{bmatrix} 1 \\ -e^{\jmath\omega\tau_0} \end{bmatrix}, \tag{3.21}$$

from which we deduce the equivalent filter:

$$\mathbf{h}'(\omega) = \frac{1}{\omega} \begin{bmatrix} 1 \\ -e^{\jmath\omega\tau_0} \end{bmatrix}, \tag{3.22}$$

since $\frac{\mathcal{C}}{2} = \jmath \frac{1}{2\tau_0}$ is a constant across frequencies. Taking the complex conjugate of the components of $\mathbf{h}'(\omega)$, we get the gains that should be applied at the two microphone outputs:

$$H_1'^*(\omega) = \frac{1}{\omega}, \tag{3.23}$$

$$H_2'^*(\omega) = -\frac{1}{\omega} e^{-\jmath\omega\tau_0}, \tag{3.24}$$

where the term $e^{-\jmath\omega\tau_0}$ is simply a time-domain delay τ_0 at the second microphone output. Figure 3.6 illustrates the implementation of the first-order cardioid.

The beampattern of the beamformer $\mathbf{h}'(\omega)$ is

$$\mathcal{B}[\mathbf{h}'(\omega), \theta] = \frac{1}{\omega} \left[1 - e^{\jmath\omega\tau_0(1+\cos\theta)} \right]. \tag{3.25}$$

Figure 3.7 displays the patterns from (3.25) for several frequencies and two values of δ. Applying the approximation (3.6) to (3.25) leads to

$$\mathcal{B}[\mathbf{h}'(\omega), \theta] \approx \frac{2}{\mathcal{C}} \left(\frac{1}{2} + \frac{1}{2}\cos\theta \right), \tag{3.26}$$

which is the pattern of the first-order cardioid.

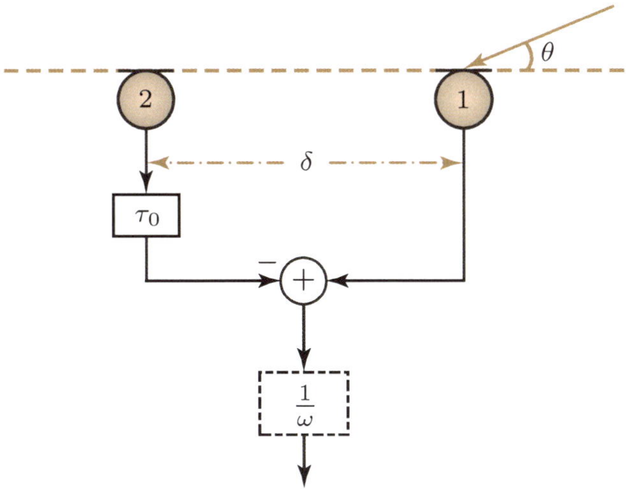

Fig. 3.6 Implementation of the first-order cardioid.

The white noise gain is

$$\begin{aligned}\mathcal{G}_{\mathrm{wn},1}\left[\mathbf{h}'(\omega)\right] &= \frac{1}{2}\left|1 - e^{\jmath 2\omega\tau_0}\right|^2 \\ &= 1 - \cos\left(2\omega\tau_0\right) \\ &= 2\left[1 - \cos\left(\omega\tau_0\right)\right]\left[1 + \cos\left(\omega\tau_0\right)\right].\end{aligned} \tag{3.27}$$

Figure 3.8 gives plots of $\mathcal{G}_{\mathrm{wn},1}\left[\mathbf{h}'(\omega)\right]$ from (3.27), as a function of frequency, for different values of δ. As explained in the previous section, the white noise can be amplified, especially at low frequencies. Indeed, for small values of $\omega\tau_0$, we have

$$\mathcal{G}_{\mathrm{wn},1}\left[\mathbf{h}'(\omega)\right] \approx 2\left(\omega\tau_0\right)^2 \tag{3.28}$$

and the white noise is amplified if $\omega\tau_0 < \sqrt{2}/2$. We can expect amplification for a larger range of frequencies with the dipole than with the cardioid (the different figures confirm this point).

We easily compute the directivity factor:

$$\begin{aligned}\mathcal{G}_{\mathrm{dn},1}\left[\mathbf{h}'(\omega)\right] &= \frac{\left|1 - e^{\jmath 2\omega\tau_0}\right|^2}{2\left[1 - \mathrm{sinc}\left(\omega\tau_0\right)\cos\left(\omega\tau_0\right)\right]} \\ &= \frac{1 - \cos\left(2\omega\tau_0\right)}{1 - \mathrm{sinc}\left(\omega\tau_0\right)\cos\left(\omega\tau_0\right)}.\end{aligned} \tag{3.29}$$

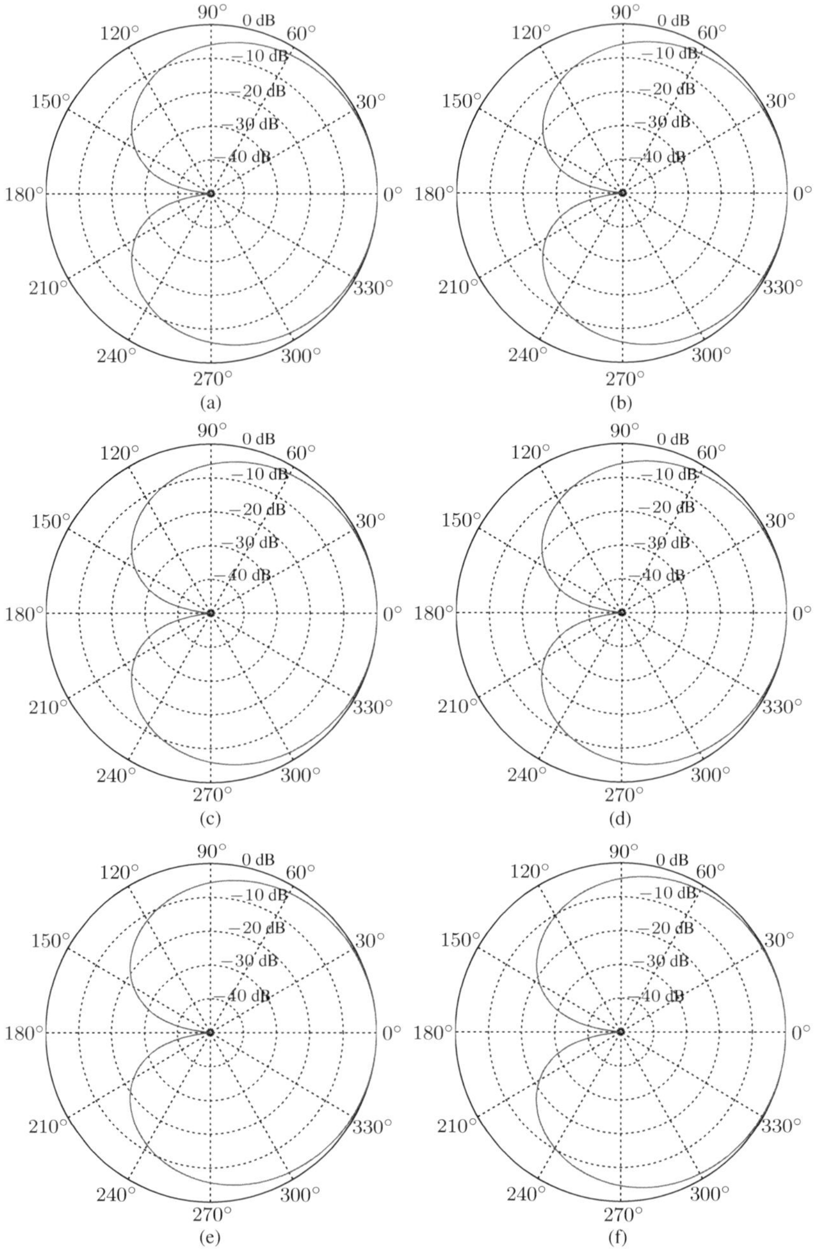

Fig. 3.7 Patterns of the first-order cardioid for several frequencies and two values of δ: (a) $f = 0.5$ kHz, $\delta = 1$ cm, (b) $f = 0.5$ kHz, $\delta = 2$ cm, (c) $f = 1$ kHz, $\delta = 1$ cm, (d) $f = 1$ kHz, $\delta = 2$ cm, (e) $f = 3$ kHz, $\delta = 1$ cm, and (f) $f = 3$ kHz, $\delta = 2$ cm.

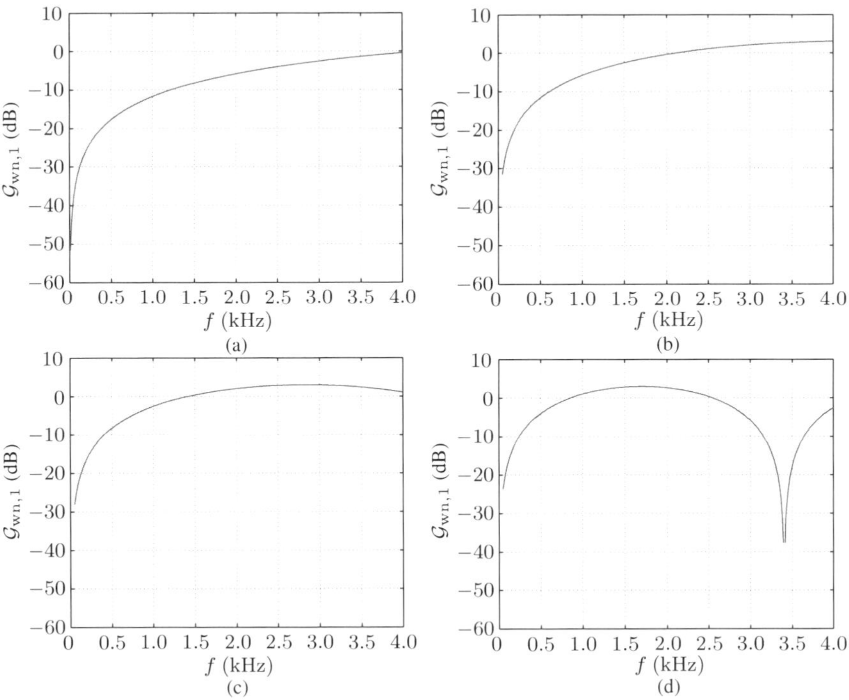

Fig. 3.8 The white noise gain of the first-order cardioid, as a function of frequency, for different values of δ: (a) $\delta = 1$ cm, (b) $\delta = 2$ cm, (c) $\delta = 3$ cm, and (d) $\delta = 5$ cm.

Figure 3.9 gives plots of $\mathcal{G}_{\text{dn},1}\left[\mathbf{h}'(\omega)\right]$ from (3.29), as a function of frequency, for different values of δ. It can be verified that

$$\mathcal{G}_{\text{dn},1}\left[\mathbf{h}'(\omega)\right] > \mathcal{G}_{\text{wn},1}\left[\mathbf{h}'(\omega)\right]. \tag{3.30}$$

For small values of $\omega\tau_0$, we have

$$\begin{aligned}\mathcal{G}_{\text{dn},1}\left[\mathbf{h}'(\omega)\right] &\approx 2\left(\omega\tau_0\right)^2 \cdot \frac{3}{2\left(\omega\tau_0\right)^2} \\ &\approx 3,\end{aligned} \tag{3.31}$$

which corresponds exactly to the theoretical value of the directivity factor for a cardioid with diffuse noise [1].

Finally, to end this section, we give the gain for a point noise source:

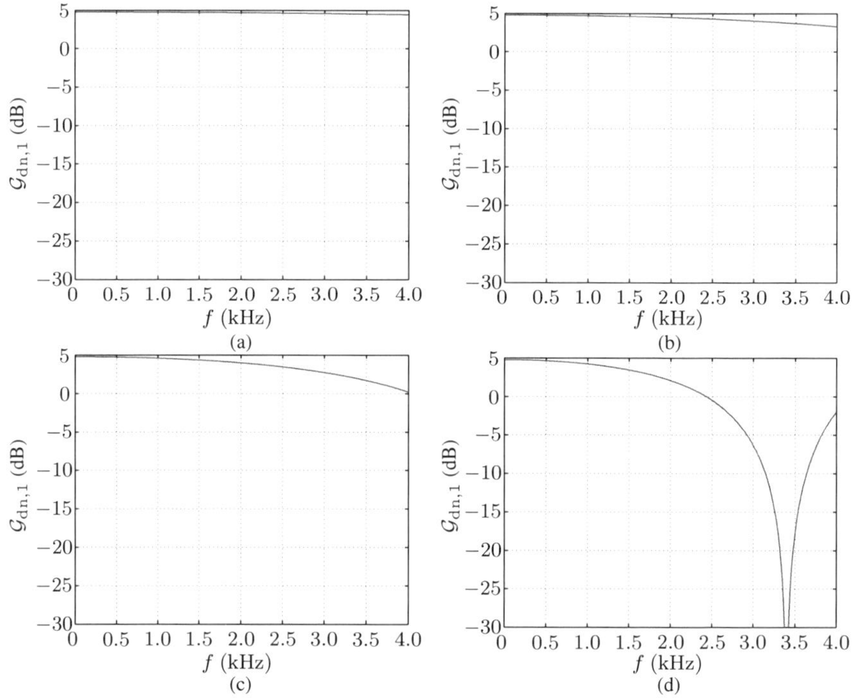

Fig. 3.9 The directivity factor of the first-order cardioid, as a function of frequency, for different values of δ: (a) $\delta = 1$ cm, (b) $\delta = 2$ cm, (c) $\delta = 3$ cm, and (d) $\delta = 5$ cm.

$$\mathcal{G}_{\rm ns,1}\left[\mathbf{h}'(\omega)\right] = \frac{\left|1 - e^{\jmath 2\omega\tau_0}\right|^2}{\left|1 - e^{\jmath\omega\tau_0(1+\cos\theta_{\rm n})}\right|^2} = \frac{1-\cos(2\omega\tau_0)}{1-\cos\left[\omega\tau_0(1+\cos\theta_{\rm n})\right]}, \tag{3.32}$$

where $\mathcal{G}_{\rm ns,1}\left[\mathbf{h}'(\omega)\right] = \infty$, $\forall f$ for $\theta_{\rm n} = 180°$. Figure 3.10 gives plots of $\mathcal{G}_{\rm ns,1}\left[\mathbf{h}'(\omega)\right]$, as a function of $\theta_{\rm n}$, for several frequencies and two values of δ. For small values of $\omega\tau_0$, (3.32) becomes

$$\mathcal{G}_{\rm ns,1}\left[\mathbf{h}'(\omega)\right] \approx \frac{1}{\left(\frac{1}{2}+\frac{1}{2}\cos\theta_{\rm n}\right)^2}, \tag{3.33}$$

which corresponds to the theoretical gain of the first-order cardioid.

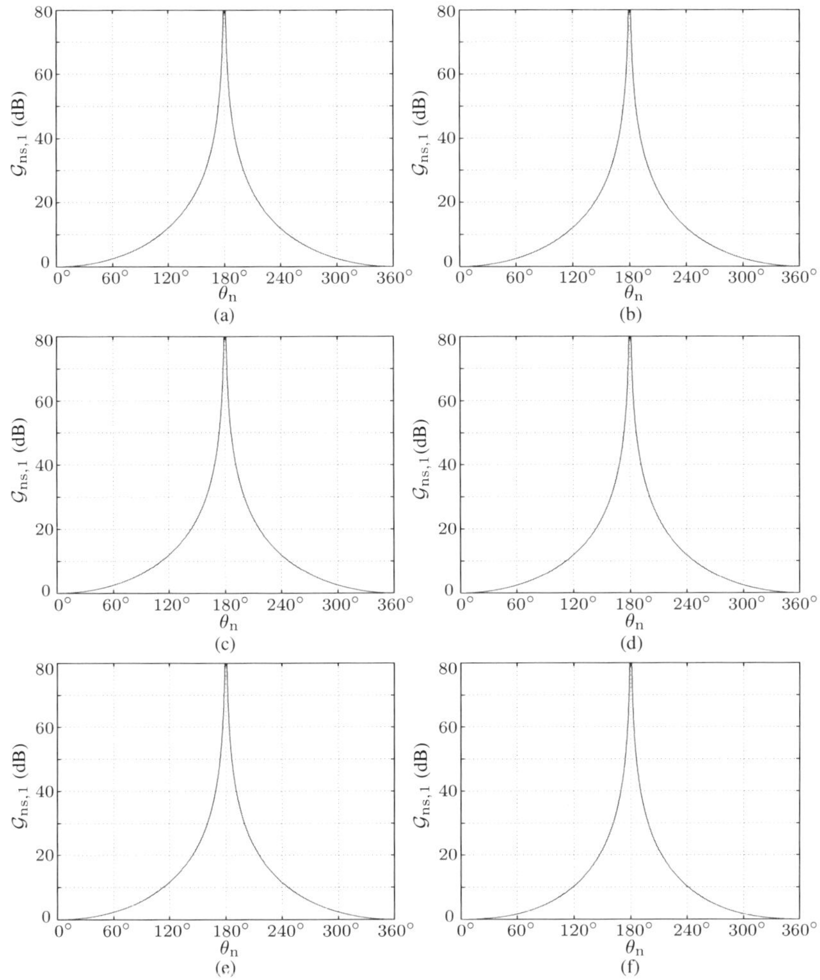

Fig. 3.10 The gain of the first-order cardioid for a point noise source, as a function of θ_{n}, for several frequencies and two values of δ: (a) $f = 0.5$ kHz, $\delta = 1$ cm, (b) $f = 0.5$ kHz, $\delta = 2$ cm, (c) $f = 1$ kHz, $\delta = 1$ cm, (d) $f = 1$ kHz, $\delta = 2$ cm, (e) $f = 3$ kHz, $\delta = 1$ cm, and (f) $f = 3$ kHz, $\delta = 2$ cm.

3.4 General First-Order Differential Array

We consider designing any first-order differential array with the linear system of two equations:

$$\begin{bmatrix} 1 & e^{\jmath\omega\tau_0} \\ 1 & e^{\jmath\omega\tau_0\alpha_{1,1}} \end{bmatrix} \mathbf{h}(\omega) = \begin{bmatrix} 1 \\ \beta_{1,1} \end{bmatrix}, \tag{3.34}$$

where $-1 \leq \alpha_{1,1} < 1$ and $0 \leq \beta_{1,1} \leq 1$. We immediately find that the unique solution to (3.34) is

$$\mathbf{h}(\omega) = \frac{1}{1 - e^{\jmath\omega\tau_0(1-\alpha_{1,1})}} \begin{bmatrix} 1 - \beta_{1,1} e^{\jmath\omega\tau_1} \\ -(1-\beta_{1,1}) e^{-\jmath\omega\tau_2} \end{bmatrix}, \tag{3.35}$$

where $\tau_1 = \tau_0(1-\alpha_{1,1})$ and $\tau_2 = \tau_0\alpha_{1,1}$. Approximating $1 - e^{\jmath\omega\tau_0(1-\alpha_{1,1})}$ with (3.6), we get

$$\mathbf{h}(\omega) \approx \frac{\mathcal{C}}{1-\alpha_{1,1}} \cdot \frac{1}{\omega} \begin{bmatrix} 1 - \beta_{1,1} e^{\jmath\omega\tau_1} \\ -(1-\beta_{1,1}) e^{-\jmath\omega\tau_2} \end{bmatrix}, \tag{3.36}$$

from which we deduce the equivalent filter[1]:

$$\mathbf{h}'(\omega) = \frac{1}{\omega} \begin{bmatrix} 1 - \beta_{1,1} e^{\jmath\omega\tau_1} \\ -(1-\beta_{1,1}) e^{-\jmath\omega\tau_2} \end{bmatrix}. \tag{3.37}$$

Taking the complex conjugate of the components of $\mathbf{h}'(\omega)$, we get the gains that should be applied at the two microphone outputs:

$$H_1'^*(\omega) = \frac{1 - \beta_{1,1} e^{-\jmath\omega\tau_1}}{\omega}, \tag{3.38}$$

$$H_2'^*(\omega) = -\frac{(1-\beta_{1,1}) e^{\jmath\omega\tau_2}}{\omega}, \tag{3.39}$$

where the term $e^{-\jmath\omega\tau_1}$ and $e^{\jmath\omega\tau_2}$ are time-domain delays τ_1 and $-\tau_2$ at the first and second microphone outputs, respectively. Figure 3.11 illustrates the implementation of the first-order differential array.

The beampattern corresponding to the beamformer $\mathbf{h}'(\omega)$ is

$$\mathcal{B}[\mathbf{h}'(\omega), \theta] = \frac{1}{\omega} \left[1 - \beta_{1,1} e^{\jmath\omega\tau_1} - (1-\beta_{1,1}) e^{\jmath\omega(\tau_0\cos\theta - \tau_2)}\right] \tag{3.40}$$

and applying the approximation (3.6) to the previous expression leads to

$$\mathcal{B}[\mathbf{h}'(\omega), \theta] \approx \frac{1-\alpha_{1,1}}{\mathcal{C}} \left[\left(1 - \frac{1-\beta_{1,1}}{1-\alpha_{1,1}}\right) + \frac{1-\beta_{1,1}}{1-\alpha_{1,1}} \cos\theta\right], \tag{3.41}$$

which is recognized to be the pattern of the first-order DMA. There are five interesting cases.

- Dipole: $\alpha_{1,1} = \beta_{1,1} = 0$ (see Section 3.2).
- Cardioid: $\alpha_{1,1} = -1$, $\beta_{1,1} = 0$ (see Section 3.3).
- Subcardioid [2], [3]: $\alpha_{1,1} = -1$, $\beta_{1,1} = 0.4$.
- Hypercardioid: $\alpha_{1,1} = -\frac{1}{2}$, $\beta_{1,1} = 0$.

[1] Notice that this filter is noncausal for $\alpha_{1,1} > 0$. In this case, we need to add a processing delay equal to $\tau_0\alpha_{1,1} = \tau_2$.

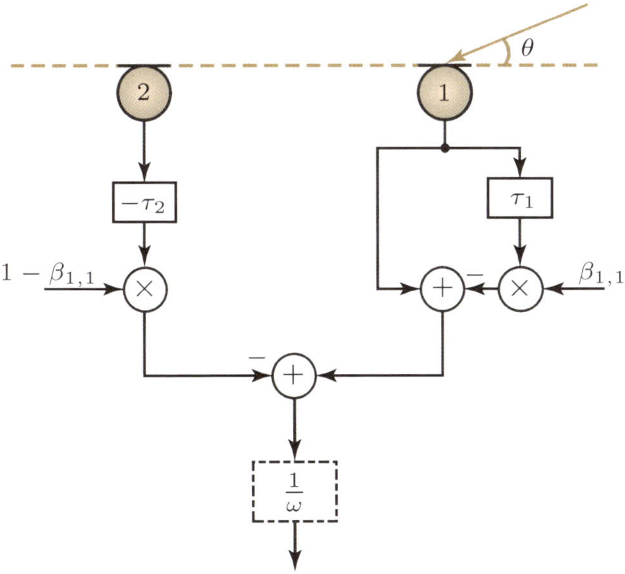

Fig. 3.11 Implementation of the first-order differential array.

- Supercardioid: $\alpha_{1,1} = \frac{1-\sqrt{2}}{2-\sqrt{2}}$, $\beta_{1,1} = 0$.

Figures 3.12, 3.13, and 3.14 display the patterns from (3.40) of the subcardioid, hypercardioid, and supercardioid, respectively, for several frequencies and two values of δ.

It can be verified that the white noise gain is

$$\mathcal{G}_{\text{wn},1}\left[\mathbf{h}'(\omega)\right] = 1 - (1-\beta_{1,1}) \frac{\cos\left[\omega\left(\tau_0 - \tau_2\right)\right] - \beta_{1,1}\cos\left[\omega\left(\tau_0 - \tau_1 - \tau_2\right)\right]}{\beta_{1,1} + \left(1-\beta_{1,1}\right)^2 - \beta_{1,1}\cos\left(\omega\tau_1\right)} \tag{3.42}$$

and for all the patterns where $\beta_{1,1} = 0$, (3.42) simplifies to

$$\begin{aligned} \mathcal{G}_{\text{wn},1}\left[\mathbf{h}'(\omega)\right] &= 1 - \cos\left[\omega\left(\tau_0 - \tau_2\right)\right] \\ &= 1 - \cos\left[\omega\tau_0\left(1 - \alpha_{1,1}\right)\right]. \end{aligned} \tag{3.43}$$

Figures 3.15 and 3.16 show plots of $\mathcal{G}_{\text{wn},1}\left[\mathbf{h}'(\omega)\right]$ from (3.43), as a function of frequency, for the hypercardioid and supercardioid, respectively, for different values of δ.

For small values of $\omega\left(\tau_0 - \tau_2\right)$, we can express (3.43) as

$$\mathcal{G}_{\text{wn},1}\left[\mathbf{h}'(\omega)\right] \approx \frac{1}{2}\left[\omega\tau_0\left(1 - \alpha_{1,1}\right)\right]^2. \tag{3.44}$$

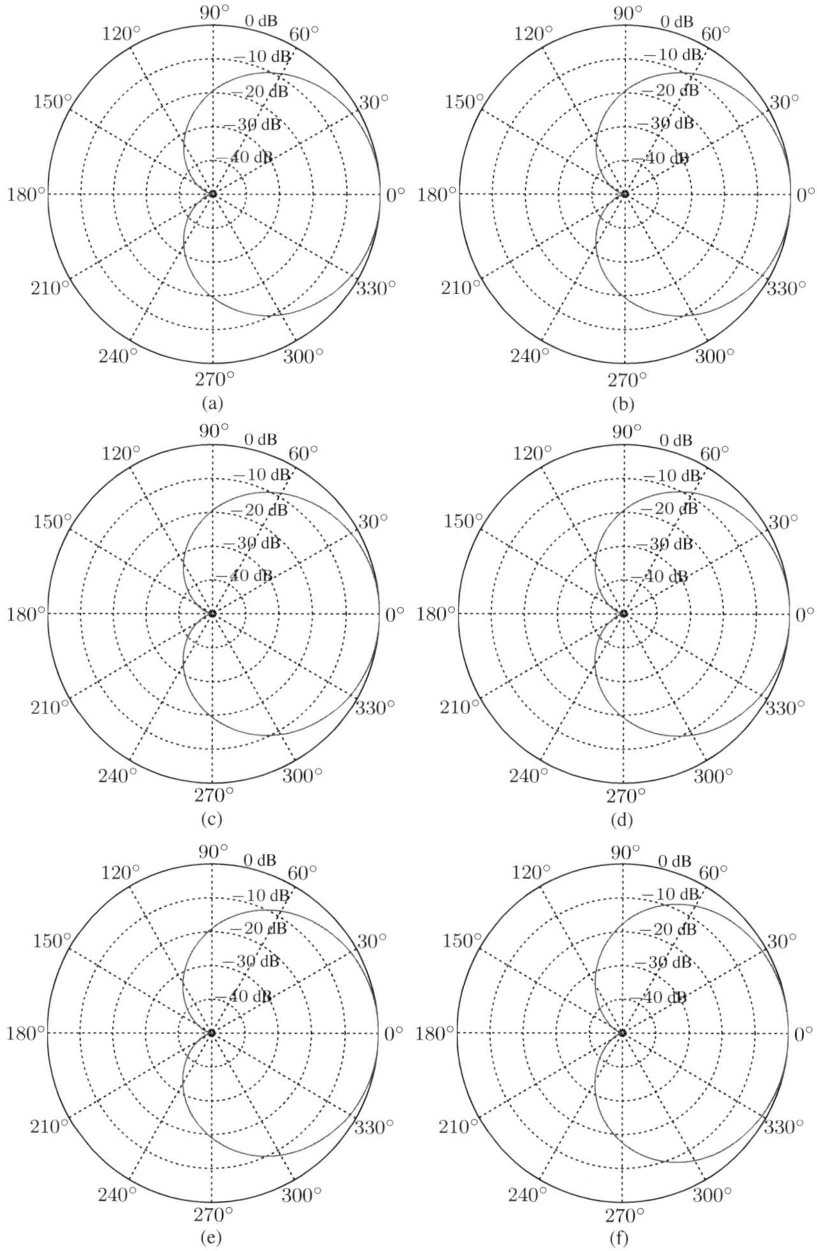

Fig. 3.12 Patterns of the first-order subcardioid for several frequencies and two values of δ: (a) $f = 0.5$ kHz, $\delta = 1$ cm, (b) $f = 0.5$ kHz, $\delta = 2$ cm, (c) $f = 1$ kHz, $\delta = 1$ cm, (d) $f = 1$ kHz, $\delta = 2$ cm, (e) $f = 3$ kHz, $\delta = 1$ cm, and (f) $f = 3$ kHz, $\delta = 2$ cm.

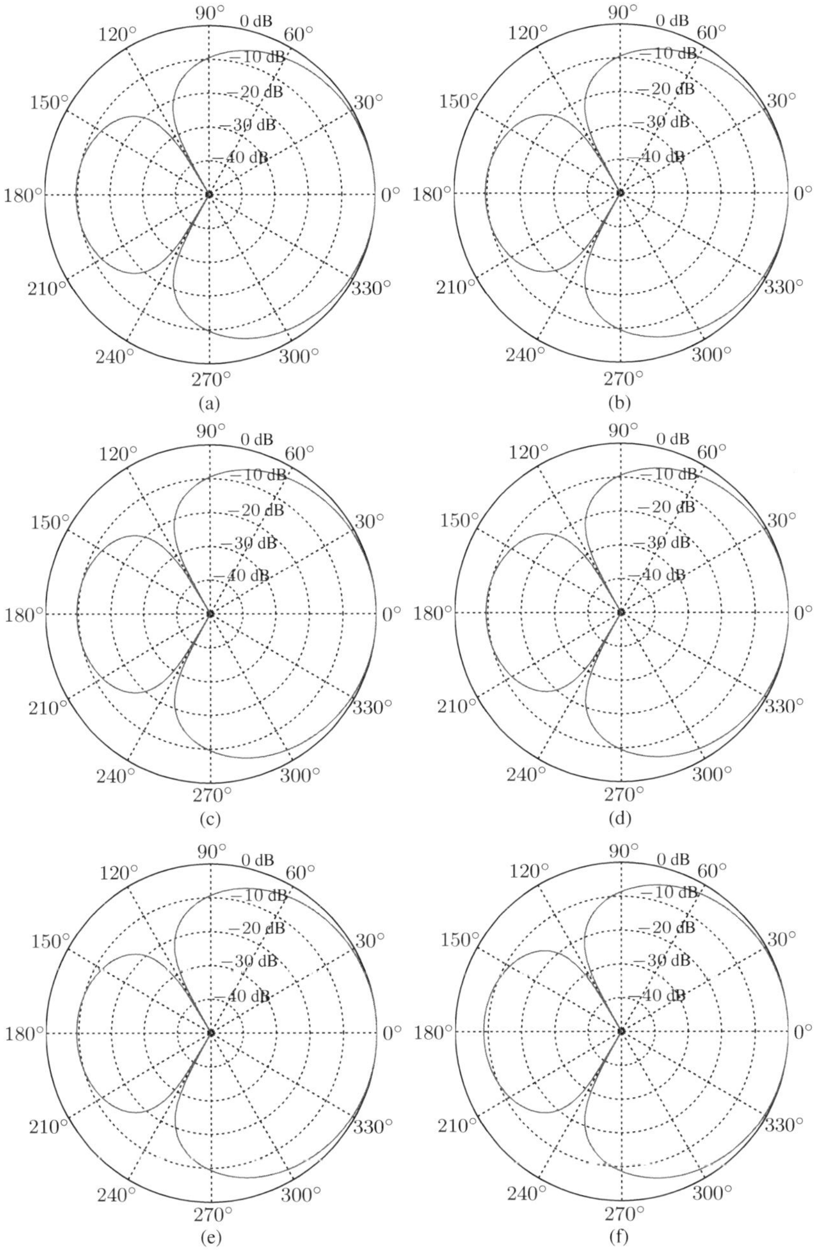

Fig. 3.13 Patterns of the first-order hypercardioid for several frequencies and two values of δ: (a) $f = 0.5$ kHz, $\delta = 1$ cm, (b) $f = 0.5$ kHz, $\delta = 2$ cm, (c) $f = 1$ kHz, $\delta = 1$ cm, (d) $f = 1$ kHz, $\delta = 2$ cm, (e) $f = 3$ kHz, $\delta = 1$ cm, and (f) $f = 3$ kHz, $\delta = 2$ cm.

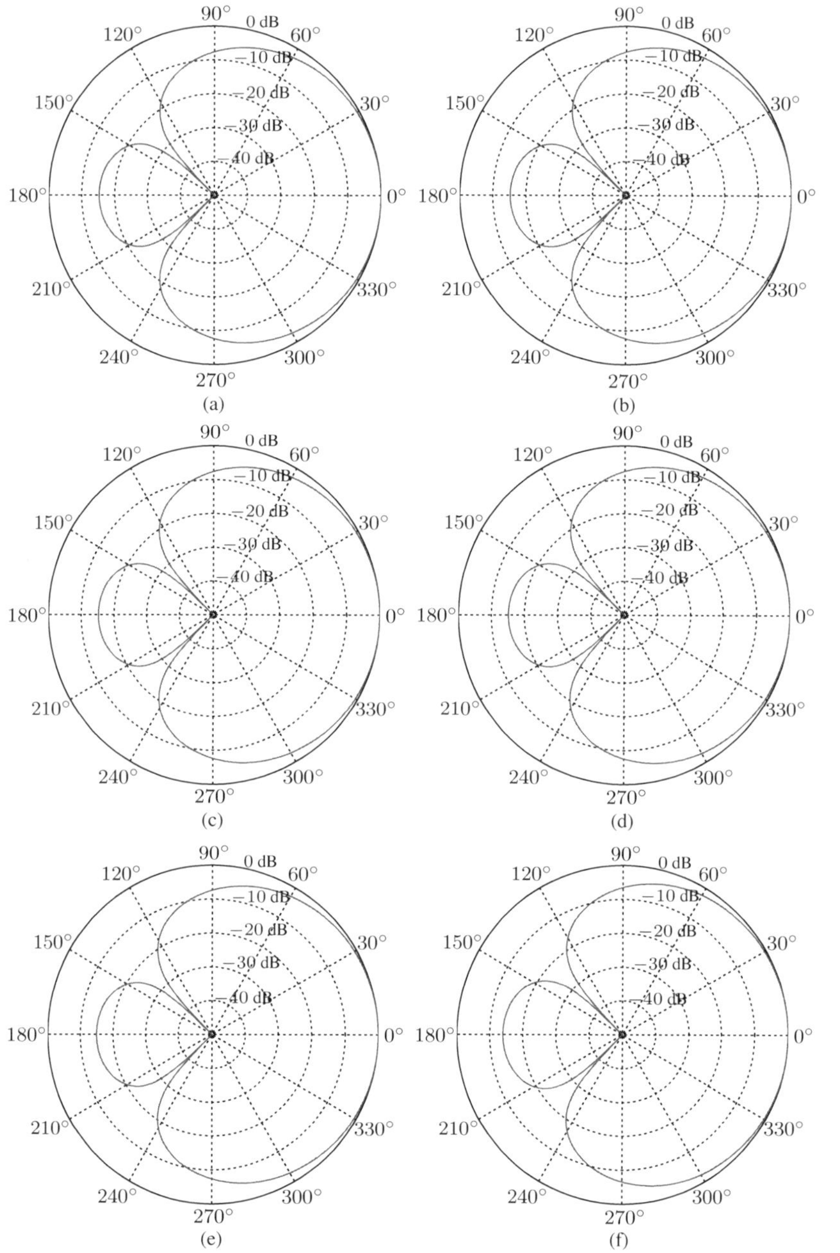

Fig. 3.14 Patterns of the first-order supercardioid for several frequencies and two values of δ: (a) $f = 0.5$ kHz, $\delta = 1$ cm, (b) $f = 0.5$ kHz, $\delta = 2$ cm, (c) $f = 1$ kHz, $\delta = 1$ cm, (d) $f = 1$ kHz, $\delta = 2$ cm, (e) $f = 3$ kHz, $\delta = 1$ cm, and (f) $f = 3$ kHz, $\delta = 2$ cm.

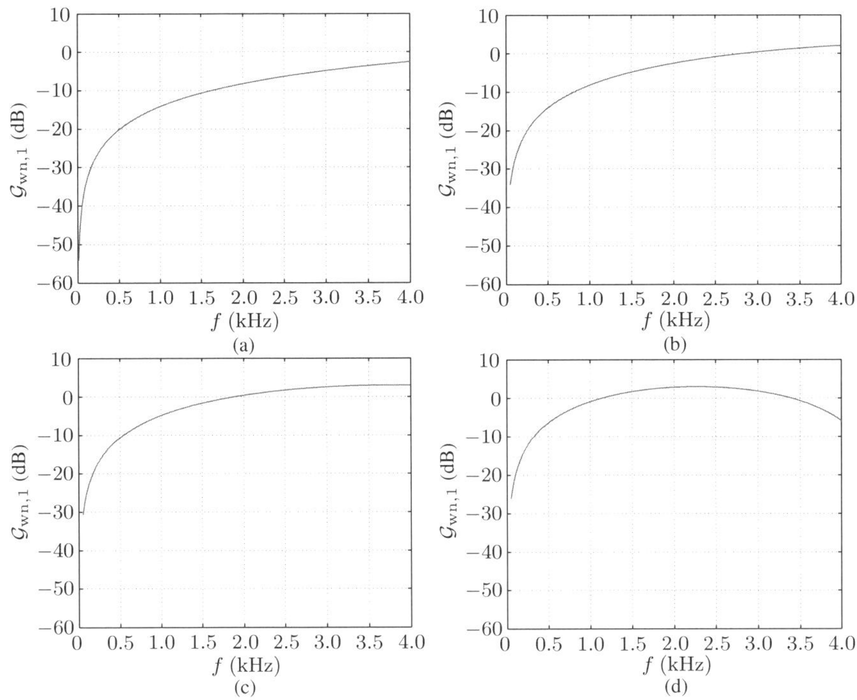

Fig. 3.15 The white noise gain of the first-order hypercardioid, as a function of frequency, for different values of δ: (a) $\delta = 1$ cm, (b) $\delta = 2$ cm, (c) $\delta = 3$ cm, and (d) $\delta = 5$ cm.

We conclude that the white noise is amplified if

$$\omega\tau_0 < \frac{\sqrt{2}}{1-\alpha_{1,1}}. \tag{3.45}$$

Interestingly, the farther away is the null from the main lobe, the smaller is the range of frequencies that can be affected by white noise amplification. This fact intuitively makes sense since by putting the null closer to the main lobe, the two constraints that shape the directional pattern will be conflicting more with each other. In other words, the linear system of two equations given in (3.34) will be ill conditioned. As a result, the uncorrelated white noise at the two microphones will be amplified. The best possible situation (where the two constraints are the farthest away and, hence, the linear system is best conditioned) is a null at $\theta = 180°$, which corresponds to the cardioid. To show more rigourously this important point, let us denote by $\mathbf{V}_2$ the 2×2 matrix involved in (3.34). The condition number of $\mathbf{V}_2$ is defined as [4]

$$\chi(\mathbf{V}_2) = \left\|\mathbf{V}_2\right\|_\infty \left\|\mathbf{V}_2^{-1}\right\|_\infty, \tag{3.46}$$

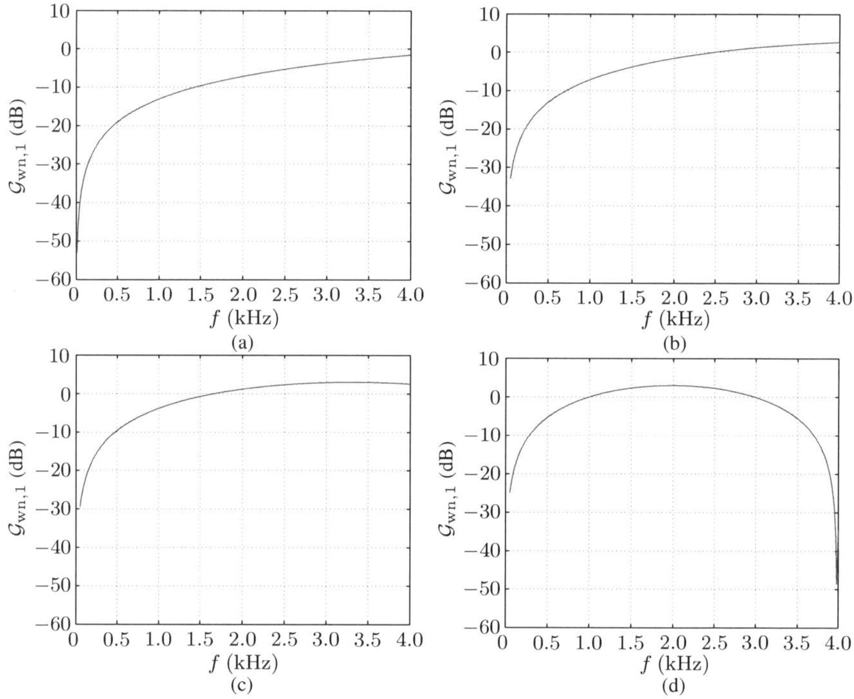

Fig. 3.16 The white noise gain of the first-order supercardioid, as a function of frequency, for different values of δ: (a) $\delta = 1$ cm, (b) $\delta = 2$ cm, (c) $\delta = 3$ cm, and (d) $\delta = 5$ cm.

where

$$\|\mathbf{M}\|_\infty = \max_{1\leq i\leq 2} \sum_{j=1}^{2} |m_{ij}| \tag{3.47}$$

is the ℓ_∞ matrix norm of $\mathbf{M}$. It is easy to check that

$$\|\mathbf{V}_2\|_\infty = 2 \tag{3.48}$$

and

$$\begin{aligned} \left\|\mathbf{V}_2^{-1}\right\|_\infty &= \frac{2}{\left|1 - e^{\jmath\omega\tau_0\left(1-\alpha_{1,1}\right)}\right|} \\ &\approx \frac{2}{\omega\tau_0\left(1-\alpha_{1,1}\right)}. \end{aligned} \tag{3.49}$$

Therefore,

$$\chi(\mathbf{V}_2) \approx \frac{4}{\omega\tau_0(1-\alpha_{1,1})}. \tag{3.50}$$

We see from (3.50) that in the range where $\alpha_{1,1}$ is defined, the condition number is minimized for $\alpha_{1,1} = -1$, i.e., the cardioid. When $\alpha_{1,1}$ gets closer to 1, $\chi(\mathbf{V}_2)$ gets larger, which means that $\mathbf{V}_2$ is not well conditioned and white noise amplification should be expected. One easy way to improve this issue is to regularize $\mathbf{V}_2$ but this may affect the shape of the beampattern. This method is equivalent to optimizing the array gain under some constraint on the white noise gain [5].

For $\beta_{1,1} = 0$, we find that the directivity factor is

$$\begin{aligned} \mathcal{G}_{\mathrm{dn},1}[\mathbf{h}'(\omega)] &= \frac{1-\cos[\omega(\tau_0-\tau_2)]}{1-\mathrm{sinc}(\omega\tau_0)\cos(\omega\tau_2)} \\ &= \frac{1-\cos[\omega\tau_0(1-\alpha_{1,1})]}{1-\mathrm{sinc}(\omega\tau_0)\cos(\omega\tau_0\alpha_{1,1})}. \end{aligned} \tag{3.51}$$

Figures 3.17 and 3.18 give plots of $\mathcal{G}_{\mathrm{dn},1}[\mathbf{h}'(\omega)]$ from (3.51), as a function of frequency, for the hypercardioid and supercardioid, respectively, for different values of δ.

For small values of $\omega\tau_0(1-\alpha_{1,1})$, we get

$$\begin{aligned} \mathcal{G}_{\mathrm{dn},1}[\mathbf{h}'(\omega)] &\approx \frac{(\omega\tau_0)^2(1-\alpha_{1,1})^2}{2} \cdot \frac{6}{(\omega\tau_0)^2(3\alpha_{1,1}^2+1)} \\ &\approx 3\frac{(1-\alpha_{1,1})^2}{3\alpha_{1,1}^2+1}. \end{aligned} \tag{3.52}$$

This result corresponds to the theoretical value of the directivity factor for diffuse noise. For $\alpha_{1,1} = -1$, we get the directivity factor (equal to 3) of the cardioid (see Section 3.3) and for $\alpha_{1,1} = 0$, we get the directivity factor (also equal to 3) of the dipole (see Section 3.2). It is well known that another type of hypercardioid[2] is obtained by maximizing the gain with diffuse noise. According to [1], this gain is exactly equal to $4 = M^2$. The value of $\alpha_{1,1}$ corresponding to this hypercardioid is $-1/3$ and the beampattern is

$$\mathcal{B}[\mathbf{h}'(\omega),\theta] \approx \frac{4}{3\mathcal{C}}\left[\frac{1}{4}+\frac{3}{4}\cos\theta\right]. \tag{3.53}$$

Substituting $\alpha_{1,1} = -\frac{1}{3}$ into (3.52), we find that indeed

$$\mathcal{G}_{\mathrm{dn},1}[\mathbf{h}'(\omega)] \approx 4. \tag{3.54}$$

[2] We recall that the hypercardioid from Chapter 2 was derived by maximizing the gain in the presence of cylindrically isotropic noise. Unless stated otherwise, this is our definition of hypercardioid.

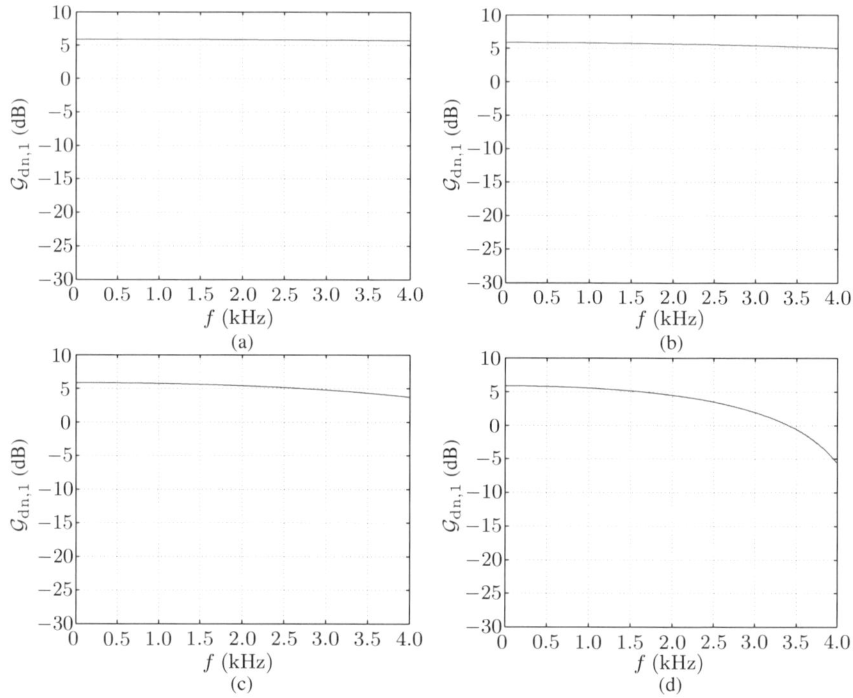

Fig. 3.17 The directivity factor of the first-order hypercardioid, as a function of frequency, for different values of δ: (a) $\delta = 1$ cm, (b) $\delta = 2$ cm, (c) $\delta = 3$ cm, and (d) $\delta = 5$ cm.

It can also be easily verified that $\mathcal{G}_{\mathrm{dn},1}\left[\mathbf{h}'(\omega)\right]$ is maximized with $\alpha_{1,1} = -\frac{1}{3}$. For our hypercardioid ($\alpha_{1,1} = -1/2$) and the supercardioid, the gains are, respectively, 27/7 and $3/\left(15 - 10\sqrt{2}\right)$. Figure 3.19 shows a plot of $\mathcal{G}_{\mathrm{dn},1}\left[\mathbf{h}'(\omega)\right]$ from (3.52), as a function of $\alpha_{1,1}$.

For $\beta_{1,1} = 0$, the gain for a point noise source is

$$\mathcal{G}_{\mathrm{ns},1}\left[\mathbf{h}'(\omega)\right] = \frac{1 - \cos\left[\omega\tau_0\left(1 - \alpha_{1,1}\right)\right]}{1 - \cos\left[\omega\tau_0\left(\cos\theta_{\mathrm{n}} - \alpha_{1,1}\right)\right]}, \tag{3.55}$$

where $\mathcal{G}_{\mathrm{ns},1}\left[\mathbf{h}'(\omega)\right] = \infty$, $\forall f$ for $\cos\theta_{\mathrm{n}} = \alpha_{1,1}$. Figures 3.20 and 3.21 show plots of $\mathcal{G}_{\mathrm{ns},1}\left[\mathbf{h}'(\omega)\right]$, as a function of θ_{n}, for the hypercardioid and supercardioid, respectively, for several frequencies and two values of δ. For small values of $\omega\tau_0$, (3.55) becomes

$$\mathcal{G}_{\mathrm{ns},1}\left[\mathbf{h}'(\omega)\right] \approx \frac{1}{\left(\frac{\alpha_{1,1}}{1 - \alpha_{1,1}} + \frac{1}{1 - \alpha_{1,1}}\cos\theta_{\mathrm{n}}\right)^2}, \tag{3.56}$$

which corresponds to the theoretical gain of the first-order DMA.

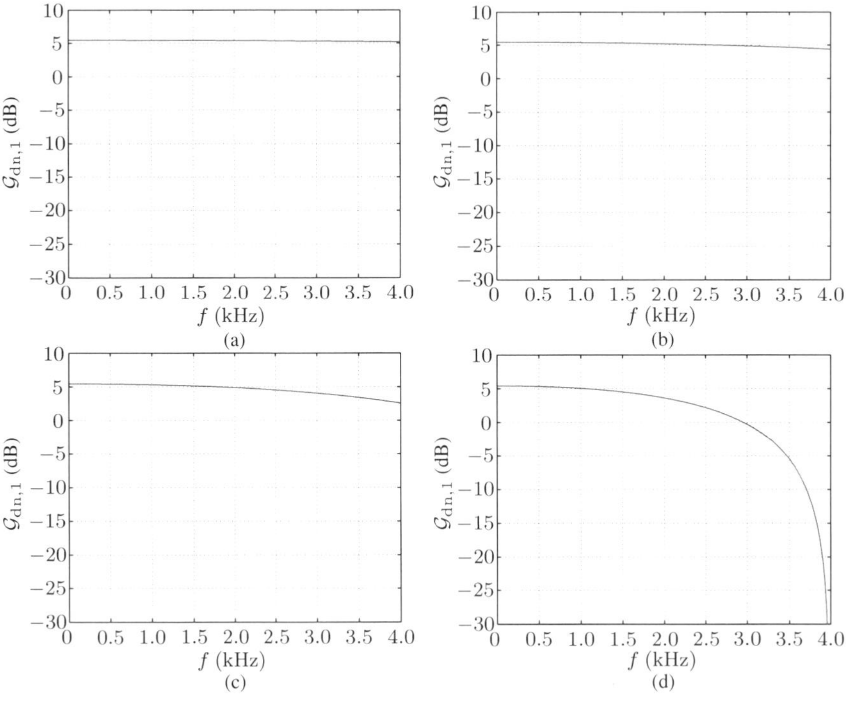

Fig. 3.18 The directivity factor of the first-order supercardioid, as a function of frequency, for different values of δ: (a) $\delta = 1$ cm, (b) $\delta = 2$ cm, (c) $\delta = 3$ cm, and (d) $\delta = 5$ cm.

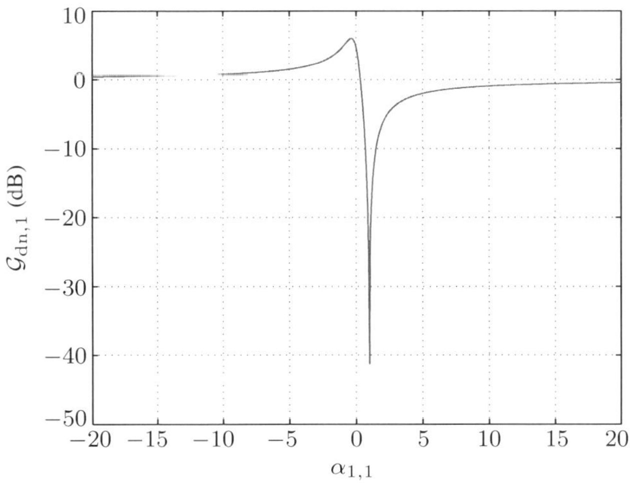

Fig. 3.19 The theoretical directivity factor of the first-order DMA as a function of $\alpha_{1,1}$.

3.5 Steering

It is very well known that electronic steering with a differential uniform linear

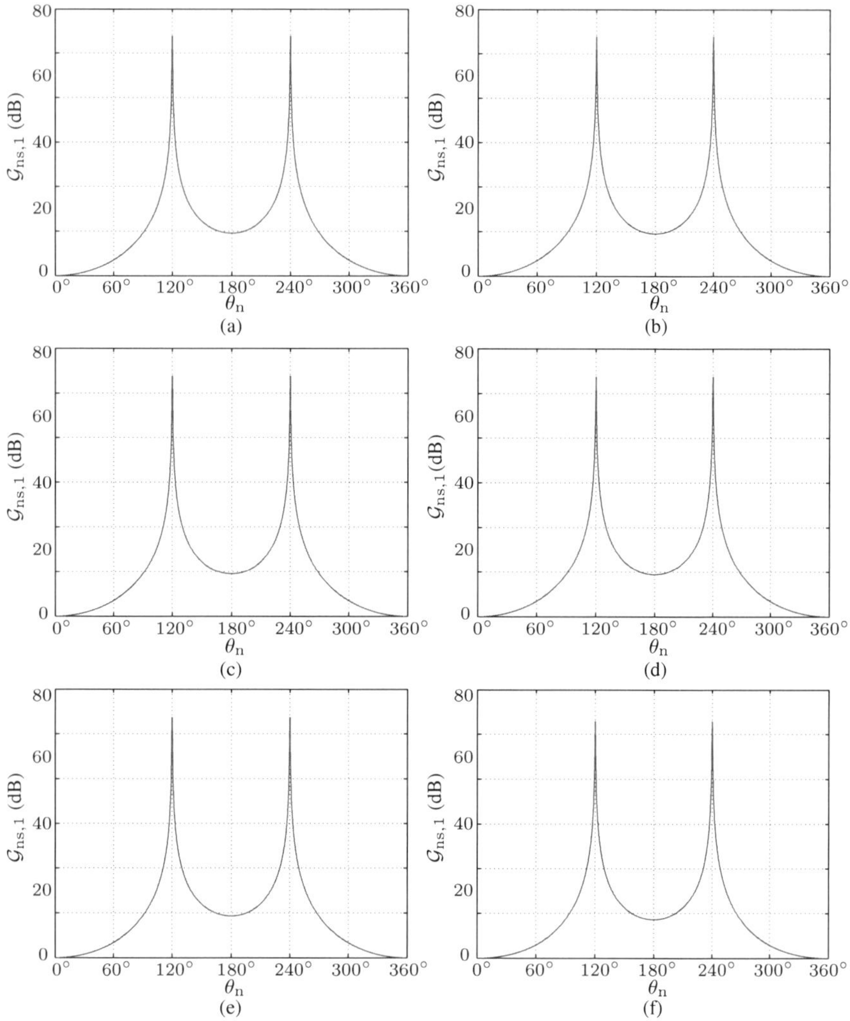

Fig. 3.20 The gain of the first-order hypercardioid for a point noise source, as a function of θ_n, for several frequencies and two values of δ: (a) $\omega = 0.125\pi$, $\delta = 1$ cm, (b) $\omega = 0.125\pi$, $\delta = 2$ cm, (c) $\omega = 0.5\pi$, $\delta = 1$ cm, (d) $\omega = 0.5\pi$, $\delta = 2$ cm, (e) $\omega = \pi$, $\delta = 1$ cm, and (f) $\omega = \pi$, $\delta = 2$ cm.

we can only form any directional patterns at the angles 0° and 180°. This problem is due to the fact that the beampattern of the beamformer $\mathbf{h}'(\omega)$ [or the steering vector, $\mathbf{d}(\omega, \cos\theta)$] is an even function of the variable θ. As a result, the main lobe of a directional pattern at a direction different from 0° (or 180°) will be affected because of the symmetry about the axis 0° – 180° on a polar plot.

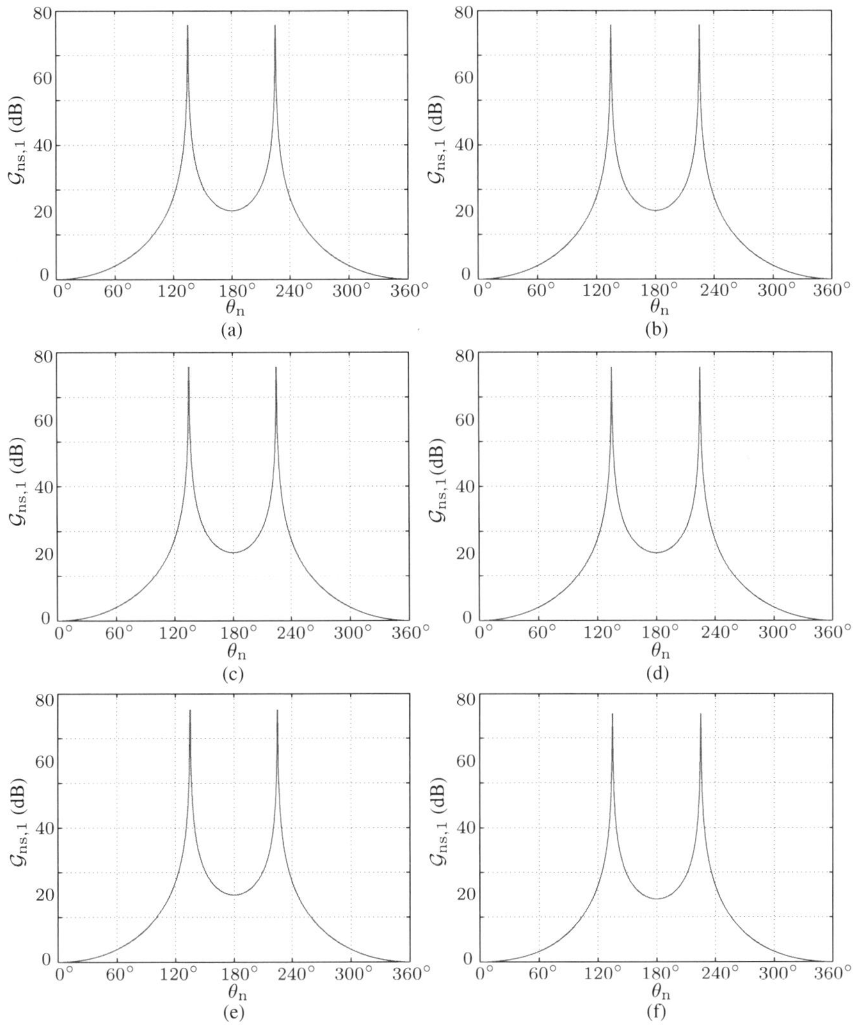

Fig. 3.21 The gain of the first-order supercardioid for a point noise source, as a function of $\theta_{\rm n}$, for several frequencies and two values of δ: (a) $\omega = 0.125\pi$, $\delta = 1$ cm, (b) $\omega = 0.125\pi$, $\delta = 2$ cm, (c) $\omega = 0.5\pi$, $\delta = 1$ cm, (d) $\omega = 0.5\pi$, $\delta = 2$ cm, (e) $\omega = \pi$, $\delta = 1$ cm, and (f) $\omega = \pi$, $\delta = 2$ cm.

Let us denote by $\theta_{\rm s}$, the steering angle. The beampattern of the Nth-order DMA is now

$$\mathcal{B}_N(\theta, \theta_{\rm s}) = \sum_{n=0}^{N} a_{N,n} \cos^n(\theta - \theta_{\rm s}). \tag{3.57}$$

The only way to be able to match (3.57) with a beamformer $\mathbf{h}'(\omega)$ is to have a steering vector of the form:

$$\mathbf{d}(\omega,\theta) = \left[e^{\jmath\omega\tau_0 \cos(\theta - \psi_1)} \; e^{\jmath\omega\tau_0 \cos(\theta - \psi_2)} \; \dots \; e^{\jmath\omega\tau_0 \cos(\theta - \psi_M)} \right]^T, \tag{3.58}$$

where ψ_m is an angle depending on the mth microphone position. This implies that, in order to be able to steer anywhere in a plane, we need at least three noncolinear microphones.

One simple way to have some (limited) discrete steering without any complicated signal processing, is to add one microphone and form an array having the shape of an equilateral triangle. With this structure, we can form three distinguished pairs of microphones and we see that we can realize all first-order directional patterns at six different directions: $0°$, $60°$, $120°$, $180°$, $240°$, and $300°$.

References

1. G. W. Elko, "Superdirectional microphone arrays," in *Acoustic Signal Processing for Telecommunication*, S. L. Gay and J. Benesty, Eds. Boston, MA: Kluwer Academic Publishers, 2000, Chapter 10, pp. 181–237.
2. J. Merimaa, "Applications of 3-D microphone array," in *Proc. AES 112th Convention*, 2002, pp. 1–11.
3. E. De Sena, H. Hacihabiboğlu, and Z. Cvetković, "On the design and implementation of higher-order differential microphones," *IEEE Trans. Audio, Speech, Language Process.*, vol. 20, pp. 162–174, Jan. 2012.
4. G. H. Golub and C. F. Van Loan, *Matrix Computations*. Third Edition. Baltimore, Maryland: The Johns Hopkins University Press, 1996.
5. H. Cox, R. M. Zeskind, and M. M. Owen, "Robust adaptive beamforming," *IEEE Trans. Acoust., Speech, Signal Process.*, vol. ASSP-35, pp. 1365–1376, Oct. 1987.
6. G. W. Elko and A.-T. Nguyen Pong, "A steerable and variable first-order differential microphone array," in *Proc. IEEE ICASSP*, 1997, pp. 223–226.
7. R. M. M. Derkx and K. Janse, "Theoretical analysis of a first-order azimuth-steerable superdirective microphone array," *IEEE Trans. Audio, Speech, Language Process.*, vol. 17, pp. 150–162, Jan. 2011.

Chapter 4
Study and Design of Second-Order Differential Arrays

In this chapter, we study and design the most important second-order differential microphone arrays. We are concerned with the dipole, two types of cardioid, quadrupole, hypercardioid, and supercardioid.

4.1 Principle

Any second-order DMA can be realized with three microphones. Therefore, we assume that we have exactly three microphones. As a result, we have three constraints to fulfill with the first one being, as usual, a one at the angle $\theta = 0°$. We deduce that the general linear system of equations to design any second-order differential array is

$$\begin{bmatrix} \mathbf{d}^H(\omega, 1) \\ \mathbf{d}^H(\omega, \alpha_{2,1}) \\ \mathbf{d}^H(\omega, \alpha_{2,2}) \end{bmatrix} \mathbf{h}(\omega) = \begin{bmatrix} 1 \\ \beta_{2,1} \\ \beta_{2,2} \end{bmatrix}, \tag{4.1}$$

where $-1 \leq \alpha_{2,1} < 1$, $-1 \leq \alpha_{2,2} < 1$, $\alpha_{2,1} \neq \alpha_{2,2}$, $0 \leq \beta_{2,1} \leq 1$, and $0 \leq \beta_{2,2} \leq 1$.

Let us denote by

$$\begin{aligned} \mathbf{V}_3 &= \begin{bmatrix} \mathbf{d}^H(\omega, 1) \\ \mathbf{d}^H(\omega, \alpha_{2,1}) \\ \mathbf{d}^H(\omega, \alpha_{2,2}) \end{bmatrix} \\ &= \begin{bmatrix} 1 & v_1 & v_1^2 \\ 1 & v_2 & v_2^2 \\ 1 & v_3 & v_3^2 \end{bmatrix} \end{aligned} \tag{4.2}$$

the 3×3 Vandermonde matrix that appears in (4.1), where $v_1 = e^{\jmath\omega\tau_0}$, $v_2 = e^{\jmath\omega\tau_0\alpha_{2,1}}$, and $v_3 = e^{\jmath\omega\tau_0\alpha_{2,2}}$. From the decomposition $\mathbf{V}_3^{-1} = \mathbf{U}_3\mathbf{L}_3$

J. Benesty and J. Chen, *Study and Design of Differential Microphone Arrays*,
Springer Topics in Signal Processing 6, DOI: 10.1007/978-3-642-33753-6_4,

given in Chapter 2, where

$$\mathbf{U}_3 = \begin{bmatrix} 1 & -v_1 & v_1 v_2 \\ 0 & 1 & -(v_1 + v_2) \\ 0 & 0 & 1 \end{bmatrix} \tag{4.3}$$

and

$$\mathbf{L}_3 = \begin{bmatrix} 1 & 0 & 0 \\ \dfrac{1}{v_1 - v_2} & \dfrac{1}{v_2 - v_1} & 0 \\ \dfrac{1}{(v_1 - v_2)(v_1 - v_3)} & \dfrac{1}{(v_2 - v_1)(v_2 - v_3)} & \dfrac{1}{(v_3 - v_1)(v_3 - v_2)} \end{bmatrix}, \tag{4.4}$$

we find that the inverse of $\mathbf{V}_3$ is

$$\mathbf{V}_3^{-1} = \begin{bmatrix} \dfrac{v_2 v_3}{(v_2 - v_1)(v_3 - v_1)} & -\dfrac{v_1 v_3}{(v_2 - v_1)(v_3 - v_2)} & \dfrac{v_1 v_2}{(v_3 - v_1)(v_3 - v_2)} \\ -\dfrac{v_2 + v_3}{(v_2 - v_1)(v_3 - v_1)} & \dfrac{v_1 + v_3}{(v_2 - v_1)(v_3 - v_2)} & -\dfrac{v_1 + v_2}{(v_3 - v_1)(v_3 - v_2)} \\ \dfrac{1}{(v_2 - v_1)(v_3 - v_1)} & -\dfrac{1}{(v_2 - v_1)(v_3 - v_2)} & \dfrac{1}{(v_3 - v_1)(v_3 - v_2)} \end{bmatrix}. \tag{4.5}$$

This inverse will be of great help in the rest of this study to find the filter $\mathbf{h}(\omega)$ for any desired second-order DMA.

4.2 Second-Order Dipole

The dipole has a one at 0°, a null at 90°, and a one at 180°. Hence, its corresponding coefficients to solve (4.1) are $\alpha_{2,1} = 0$, $\alpha_{2,2} = -1$, $\beta_{2,1} = 0$, and $\beta_{2,1} = 1$. The solution to (4.1) is then

$$\mathbf{h}(\omega) = \frac{1}{\left(1 - e^{\jmath\omega\tau_0}\right)\left(1 - e^{\jmath 2\omega\tau_0}\right)} \begin{bmatrix} 1 \\ -1 - e^{\jmath\omega\tau_0} \\ e^{\jmath\omega\tau_0} \end{bmatrix} + \frac{1}{\left(1 - e^{-\jmath\omega\tau_0}\right)\left(1 - e^{-\jmath 2\omega\tau_0}\right)} \begin{bmatrix} 1 \\ -1 - e^{-\jmath\omega\tau_0} \\ e^{-\jmath\omega\tau_0} \end{bmatrix}. \tag{4.6}$$

Using the approximation $e^x \approx 1 + x$ in (4.6), we get

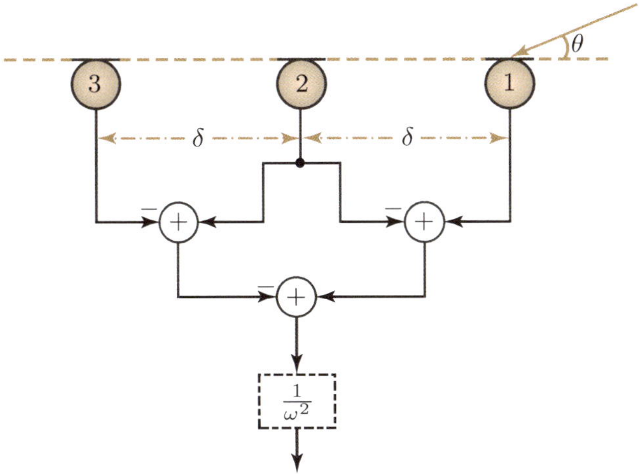

Fig. 4.1 Implementation of the second-order dipole.

$$\mathbf{h}(\omega) \approx -\frac{1}{\omega^2\tau_0^2}\begin{bmatrix} 1 \\ -1-\cos(\omega\tau_0) \\ \cos(\omega\tau_0) \end{bmatrix}$$
$$\approx -\frac{1}{\omega^2\tau_0^2}\begin{bmatrix} 1 \\ -2 \\ 1 \end{bmatrix}. \tag{4.7}$$

Since $\mathcal{C}^2 = -\frac{1}{\tau_0^2}$ is a constant across frequencies, it is more convenient to work with the equivalent filter:

$$\mathbf{h}'(\omega) = \frac{1}{\omega^2}\begin{bmatrix} 1 \\ -2 \\ 1 \end{bmatrix}, \tag{4.8}$$

where the term $1/\omega^2$ in (4.8) is simply a second-order low-pass filter. We can express (4.8) as

$$\mathbf{h}'(\omega) = \frac{1}{\omega}\left\{\frac{1}{\omega}\begin{bmatrix} 1 \\ -1 \\ 0 \end{bmatrix} - \frac{1}{\omega}\begin{bmatrix} 0 \\ 1 \\ -1 \end{bmatrix}\right\}, \tag{4.9}$$

showing that the second-order dipole can be implemented as a cascade of first-order dipoles (three in total). Figure 4.1 illustrates the implementation of the second-order dipole.

The beampattern of the beamformer $\mathbf{h}'(\omega)$ is

$$\begin{aligned}\mathcal{B}\left[\mathbf{h}'(\omega),\theta\right] &= \mathbf{d}^H(\omega,\cos\theta)\,\mathbf{h}'(\omega) \\ &= \frac{1}{\omega^2}\left(1-e^{\jmath\omega\tau_0\cos\theta}\right)^2 .\end{aligned} \tag{4.10}$$

Figure 4.2 displays the patterns of $\mathbf{h}'(\omega)$ from the above equation for several frequencies and two values of δ. For small values of $\omega\tau_0$, we can approximate (4.10) as

$$\mathcal{B}\left[\mathbf{h}'(\omega),\theta\right] \approx \frac{1}{\mathcal{C}^2}\cos^2\theta, \tag{4.11}$$

where we recognize the pattern of the second-order dipole.

The white noise gain is

$$\begin{aligned}\mathcal{G}_{\mathrm{wn},2}\left[\mathbf{h}'(\omega)\right] &= \frac{\left|\mathbf{h}'^H(\omega)\,\mathbf{d}(\omega,1)\right|^2}{\mathbf{h}'^H(\omega)\,\mathbf{h}'(\omega)} \\ &= \frac{1}{6}\left|1-e^{\jmath\omega\tau_0}\right|^4 \\ &= \frac{2}{3}\left[1-\cos(\omega\tau_0)\right]^2 .\end{aligned} \tag{4.12}$$

In Fig. 4.3, we plot $\mathcal{G}_{\mathrm{wn},2}\left[\mathbf{h}'(\omega)\right]$ from (4.12), as a function of frequency, for different values of δ. For small values of $\omega\tau_0$, we get

$$\mathcal{G}_{\mathrm{wn},2}\left[\mathbf{h}'(\omega)\right] \approx \frac{(\omega\tau_0)^4}{6}, \tag{4.13}$$

from which we deduce that the white noise is amplified if

$$\omega\tau_0 < 6^{1/4} \approx 1.57. \tag{4.14}$$

Comparing this result with the one obtained for the first-order dipole (see Chapter 3), we observe that the situation for the second-order dipole is worse.

The directivity factor is

$$\begin{aligned}\mathcal{G}_{\mathrm{dn},2}\left[\mathbf{h}'(\omega)\right] &= \frac{\left|\mathbf{h}'^H(\omega)\,\mathbf{d}(\omega,1)\right|^2}{\mathbf{h}'^H(\omega)\,\mathbf{\Gamma}_{\mathrm{dn}}(\omega)\,\mathbf{h}'(\omega)} \\ &= \frac{\left|1-e^{\jmath\omega\tau_0}\right|^4}{2\left[3-4\mathrm{sinc}(\omega\tau_0)+\mathrm{sinc}(2\omega\tau_0)\right]} \\ &= \frac{2\left[1-\cos(\omega\tau_0)\right]^2}{3-4\mathrm{sinc}(\omega\tau_0)+\mathrm{sinc}(\omega\tau_0)\cos(\omega\tau_0)} .\end{aligned} \tag{4.15}$$

In Fig. 4.4, we plot $\mathcal{G}_{\mathrm{dn},2}\left[\mathbf{h}'(\omega)\right]$ from (4.15), as a function of frequency, for different values of δ. Using Taylor's series, we can show that

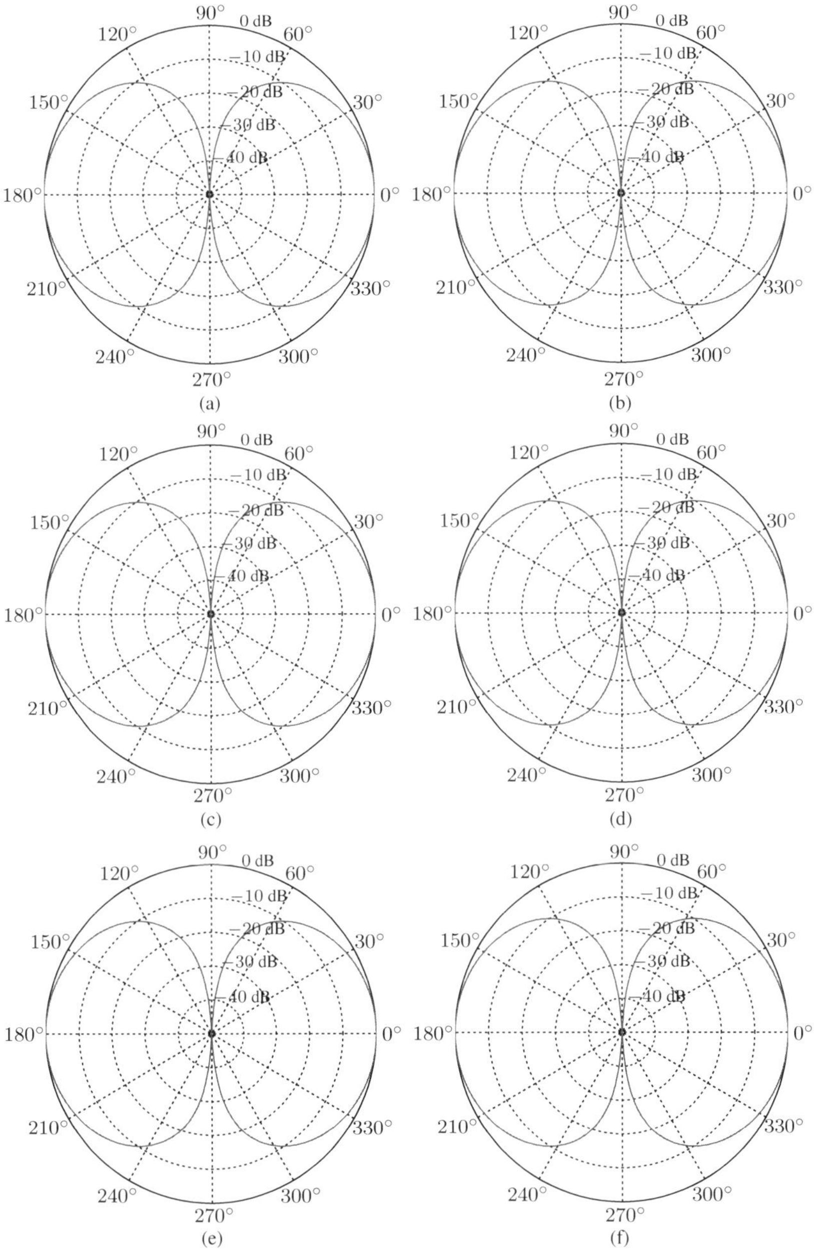

Fig. 4.2 Patterns of the second-order dipole for several frequencies and two values of δ: (a) $f = 0.5$ kHz, $\delta = 1$ cm, (b) $f = 0.5$ kHz, $\delta = 2$ cm, (c) $f = 1$ kHz, $\delta = 1$ cm, (d) $f = 1$ kHz, $\delta = 2$ cm, (e) $f = 3$ kHz, $\delta = 1$ cm, and (f) $f = 3$ kHz, $\delta = 2$ cm.

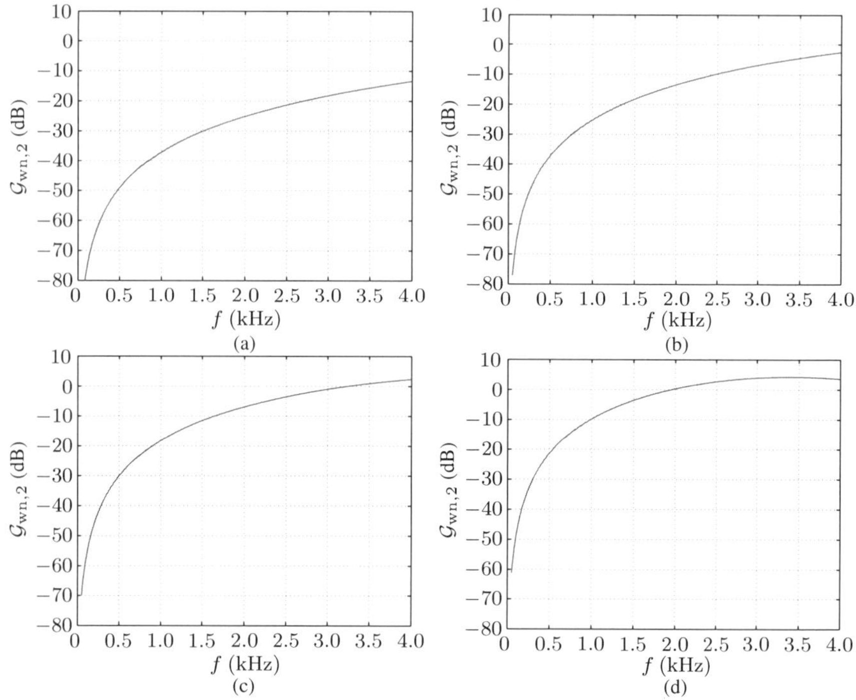

Fig. 4.3 The white noise gain of the second-order dipole, as a function of frequency, for different values of δ: (a) $\delta = 1$ cm, (b) $\delta = 2$ cm, (c) $\delta = 3$ cm, and (d) $\delta = 5$ cm.

$$\begin{aligned}\mathcal{G}_{\mathrm{dn},2}\left[\mathbf{h}'(\omega)\right] &\approx 2\frac{(\omega\tau_0)^4}{4}\cdot\frac{10}{(\omega\tau_0)^4}\\ &\approx 5,\end{aligned} \tag{4.16}$$

which should correspond to the theoretical value of the directivity factor for the second-order dipole with diffuse noise[1]. We observe that this gain is almost two times larger than the first-order dipole gain.

For a point noise source, the gain of the second-order dipole is

$$\begin{aligned}\mathcal{G}_{\mathrm{ns},2}\left[\mathbf{h}'(\omega)\right] &= \frac{\left|\mathbf{h}'^H(\omega)\mathbf{d}(\omega,1)\right|^2}{\left|\mathbf{h}'^H(\omega)\mathbf{d}(\omega,\cos\theta_{\mathrm{n}})\right|^2}\\ &= \frac{\left|1-e^{\jmath\omega\tau_0}\right|^4}{\left|1-e^{\jmath\omega\tau_0\cos\theta_{\mathrm{n}}}\right|^4}\\ &= \frac{\left[1-\cos(\omega\tau_0)\right]^2}{\left[1-\cos(\omega\tau_0\cos\theta_{\mathrm{n}})\right]^2}.\end{aligned} \tag{4.17}$$

[1] Notice that the gain derived here is identical to the one given in [1].

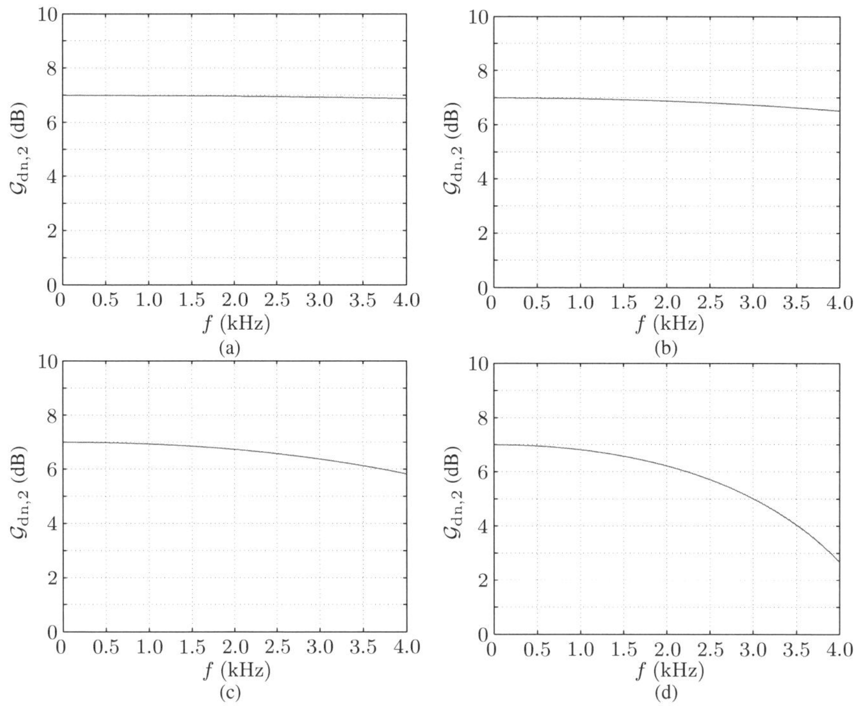

Fig. 4.4 The directivity factor of the second-order dipole, as a function of frequency, for different values of δ: (a) $\delta = 1$ cm, (b) $\delta = 2$ cm, (c) $\delta = 3$ cm, and (d) $\delta = 5$ cm.

Therefore, for $\theta_{\mathrm{n}} = 0°$ (or $180°$), $\mathcal{G}_{\mathrm{ns},2}\left[\mathbf{h}'(\omega)\right] = 1$, $\forall f$ and for $\theta_{\mathrm{n}} = 90°$ (or $270°$), $\mathcal{G}_{\mathrm{ns},2}\left[\mathbf{h}'(\omega)\right] = \infty$, $\forall f$. Figure 4.5 shows plots of $\mathcal{G}_{\mathrm{ns},2}\left[\mathbf{h}'(\omega)\right]$, as a function of θ_{n}, for several frequencies and two values of δ. With the conventional approximation, we find that

$$\mathcal{G}_{\mathrm{ns},2}\left[\mathbf{h}'(\omega)\right] \approx \frac{1}{\cos^4 \theta_{\mathrm{n}}}, \tag{4.18}$$

which corresponds to the theoretical gain of the second-order dipole.

4.3 Second-Order Cardioid, Hypercardioid, Supercardioid, and Quadrupole

The four patterns cited in the title of this section have one important thing in common: they all have two distinct nulls but in different directions. Therefore, the linear system of equations to solve in this context is

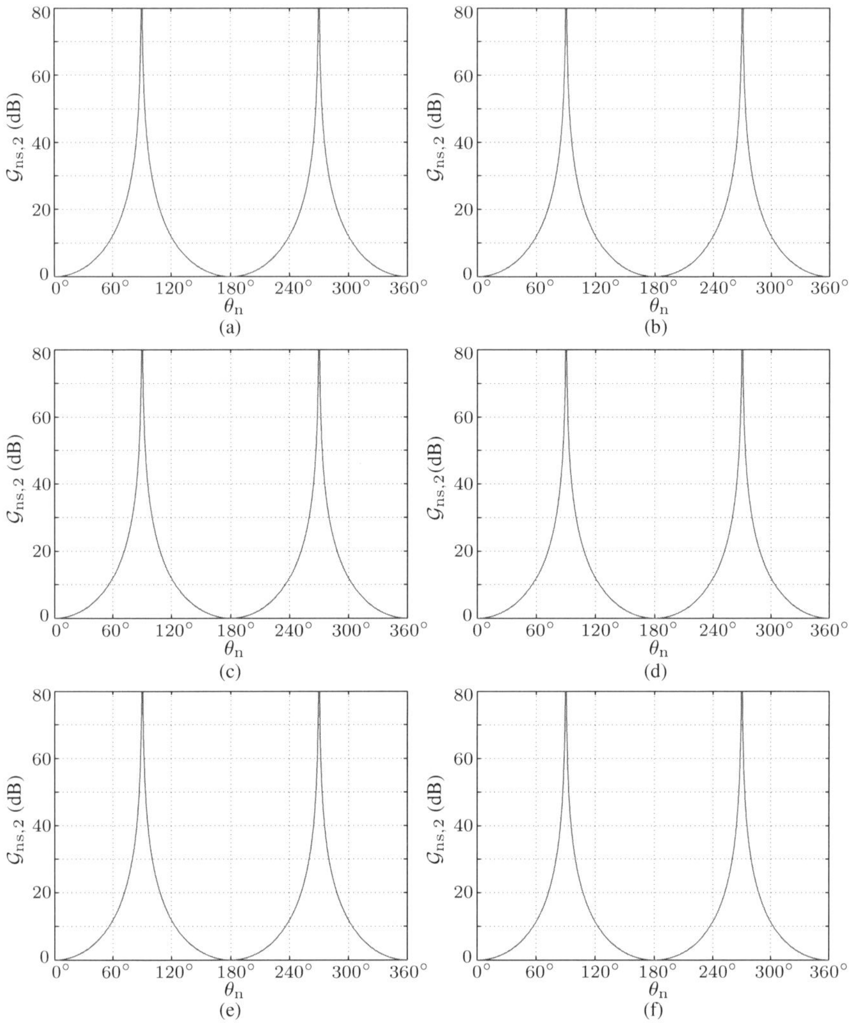

Fig. 4.5 The gain of the second-order dipole for a point noise source, as a function of $\theta_{\rm n}$, for several frequencies and two values of δ: (a) $f = 0.5$ kHz, $\delta = 1$ cm, (b) $f = 0.5$ kHz, $\delta = 2$ cm, (c) $f = 1$ kHz, $\delta = 1$ cm, (d) $f = 1$ kHz, $\delta = 2$ cm, (e) $f = 3$ kHz, $\delta = 1$ cm, and (f) $f = 3$ kHz, $\delta = 2$ cm.

$$\begin{bmatrix} \mathbf{d}^H(\omega, 1) \\ \mathbf{d}^H(\omega, \alpha_{2,1}) \\ \mathbf{d}^H(\omega, \alpha_{2,2}) \end{bmatrix} \mathbf{h}(\omega) = \begin{bmatrix} 1 \\ 0 \\ 0 \end{bmatrix}, \tag{4.19}$$

where $-1 \le \alpha_{2,1} < 1$, $-1 \le \alpha_{2,2} < 1$, and $\alpha_{2,1} \neq \alpha_{2,2}$. The values of these parameters for the different patterns of interest are as follows.

- Cardioid: $\alpha_{2,1} = -1$, $\alpha_{2,2} = 0$.

- Hypercardioid: $\alpha_{2,1} = -0.81$, $\alpha_{2,2} = 0.31$.
- Supercardioid: $\alpha_{2,1} = -0.89$, $\alpha_{2,2} = -0.28$.
- Quadrupole: $\alpha_{2,1} = -\frac{1}{\sqrt{2}}$, $\alpha_{2,2} = \frac{1}{\sqrt{2}}$.

In the rest, we assume that $\alpha_{2,1} < 0$.

It is clear that the solution to (4.19) is

$$\mathbf{h}(\omega) = \frac{1}{\left[1 - e^{\jmath\omega\tau_0(1-\alpha_{2,1})}\right]\left[1 - e^{\jmath\omega\tau_0(1-\alpha_{2,2})}\right]} \times \begin{bmatrix} 1 \\ -e^{-\jmath\omega\tau_0\alpha_{2,1}} - e^{-\jmath\omega\tau_0\alpha_{2,2}} \\ e^{-\jmath\omega\tau_0(\alpha_{2,1}+\alpha_{2,2})} \end{bmatrix}. \tag{4.20}$$

Using the conventional approximation in the factor term of (4.20), we obtain

$$\mathbf{h}(\omega) \approx \frac{c^2}{\omega^2(1-\alpha_{2,1})(1-\alpha_{2,2})} \begin{bmatrix} 1 \\ -e^{-\jmath\omega\tau_0\alpha_{2,1}} - e^{-\jmath\omega\tau_0\alpha_{2,2}} \\ e^{-\jmath\omega\tau_0(\alpha_{2,1}+\alpha_{2,2})} \end{bmatrix}, \tag{4.21}$$

from which we deduce the equivalent filter[2]:

$$\mathbf{h}'^*(\omega) = \frac{1}{\omega^2} \begin{bmatrix} 1 \\ -e^{\jmath\omega\tau_0\alpha_{2,1}} - e^{\jmath\omega\tau_0\alpha_{2,2}} \\ e^{\jmath\omega\tau_0(\alpha_{2,1}+\alpha_{2,2})} \end{bmatrix}, \tag{4.22}$$

that we can rewrite as

$$\mathbf{h}'^*(\omega) = \frac{1}{\omega}\left\{\frac{1}{\omega}\begin{bmatrix} 1 \\ -e^{\jmath\omega\tau_0\alpha_{2,1}} \\ 0 \end{bmatrix} - \frac{e^{\jmath\omega\tau_0\alpha_{2,2}}}{\omega}\begin{bmatrix} 0 \\ 1 \\ -e^{\jmath\omega\tau_0\alpha_{2,1}} \end{bmatrix}\right\}, \tag{4.23}$$

showing that the second-order DMA can be implemented as a cascade of first-order DMAs (three in total). Figure 4.6 illustrates the implementation of the second-order differential array.

It is of interest to compare the filter of the quadrupole:

$$\mathbf{h}'(\omega) = \frac{1}{\omega^2}\begin{bmatrix} 1 \\ -2\cos\left(\sqrt{2}\omega\tau_0/2\right) \\ 1 \end{bmatrix} \tag{4.24}$$

with the filter of the second-order dipole:

[2] Notice that this filter is noncausal for $\alpha_{2,2} > 0$. In this case, we need to add a processing delay equal to $\tau_0 \max(\alpha_{2,2}, \alpha_{2,1} + \alpha_{2,2})$.

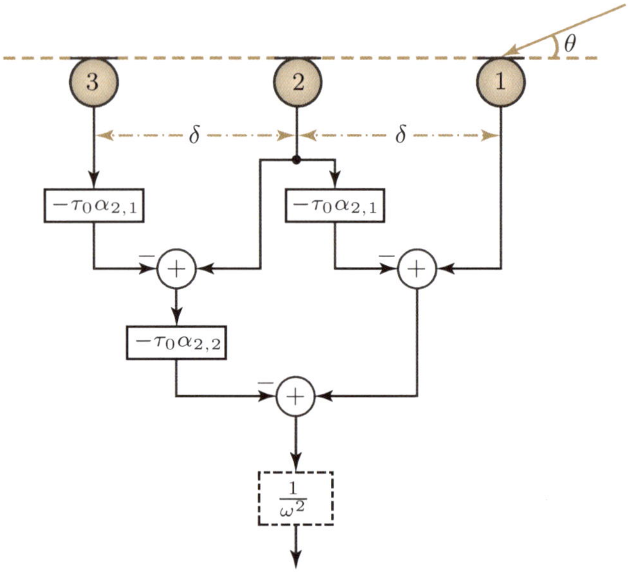

Fig. 4.6 Implementation of the second-order differential array.

$$\mathbf{h}'(\omega) = \frac{1}{\omega^2}\begin{bmatrix} 1 \\ -1-\cos(\omega\tau_0) \\ \cos(\omega\tau_0) \end{bmatrix}. \tag{4.25}$$

We recall that in the previous section, we approximated $\cos(\omega\tau_0)$ with 1; while this is possible for the dipole, it is not possible to approximate $\cos\left(\sqrt{2}\omega\tau_0/2\right)$ with 1 for the quadrupole, otherwise we get the dipole. Also, it is of interest to notice that the dipole given in (4.25) should be better than the one given in (4.8) in terms of the directional gain.

It is easy to see that the beampattern of the beamformer from (4.22) is

$$\mathcal{B}\left[\mathbf{h}'(\omega), \theta\right] = \frac{1}{\omega^2}\left[1 - e^{\jmath\omega\tau_0(\cos\theta - \alpha_{2,1})}\right]\left[1 - e^{\jmath\omega\tau_0(\cos\theta - \alpha_{2,2})}\right]. \tag{4.26}$$

Figures 4.7, 4.8, 4.9, and 4.10 display the patterns from (4.26), for the cardioid, hypercardioid, supercardioid, and quadrupole, respectively, for several frequencies and two values of δ. We can approximate (4.26) as the frequency-independent second-order directional pattern:

$$\mathcal{B}\left[\mathbf{h}'(\omega), \theta\right] \approx \frac{1}{\mathcal{C}^2}(\cos\theta - \alpha_{2,1})(\cos\theta - \alpha_{2,2}). \tag{4.27}$$

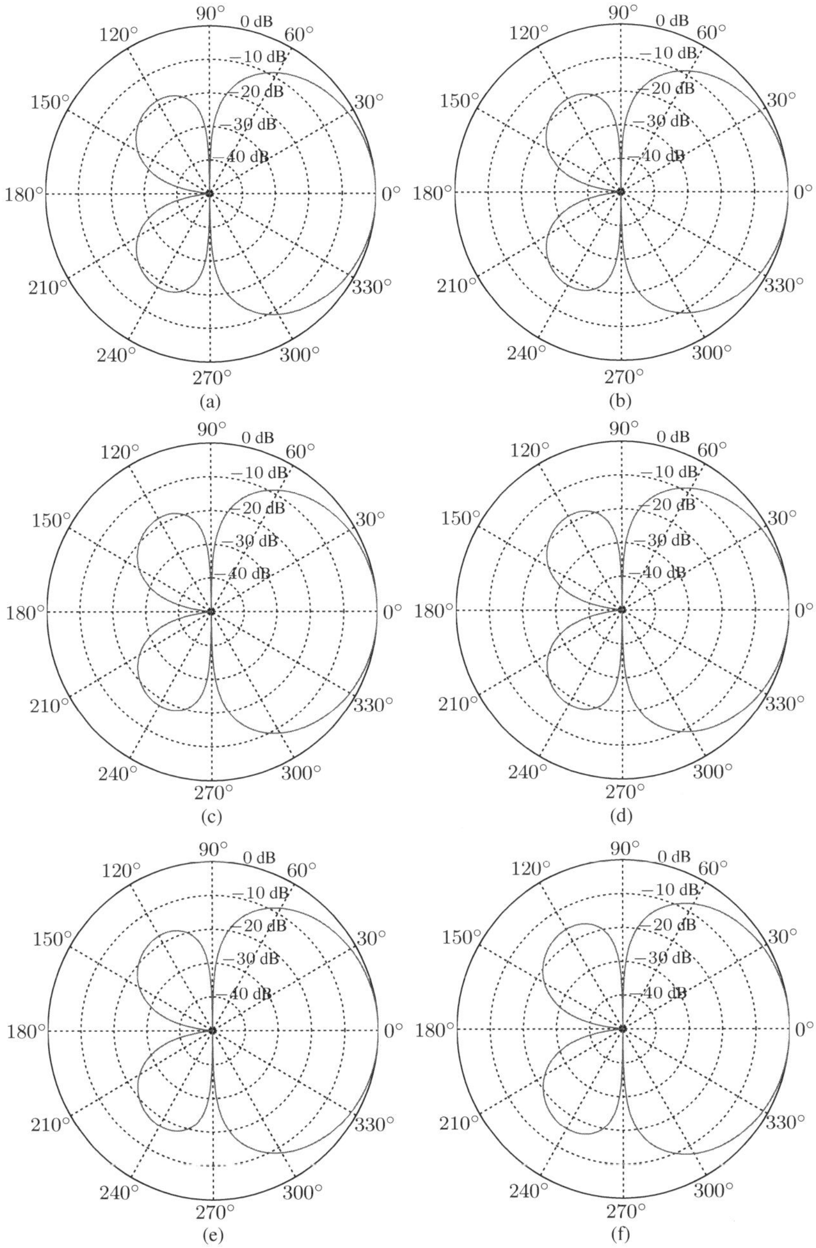

Fig. 4.7 Patterns of the second-order cardioid for several frequencies and two values of δ: (a) $f = 0.5$ kHz, $\delta = 1$ cm, (b) $f = 0.5$ kHz, $\delta = 2$ cm, (c) $f = 1$ kHz, $\delta = 1$ cm, (d) $f = 1$ kHz, $\delta = 2$ cm, (e) $f = 3$ kHz, $\delta = 1$ cm, and (f) $f = 3$ kHz, $\delta = 2$ cm.

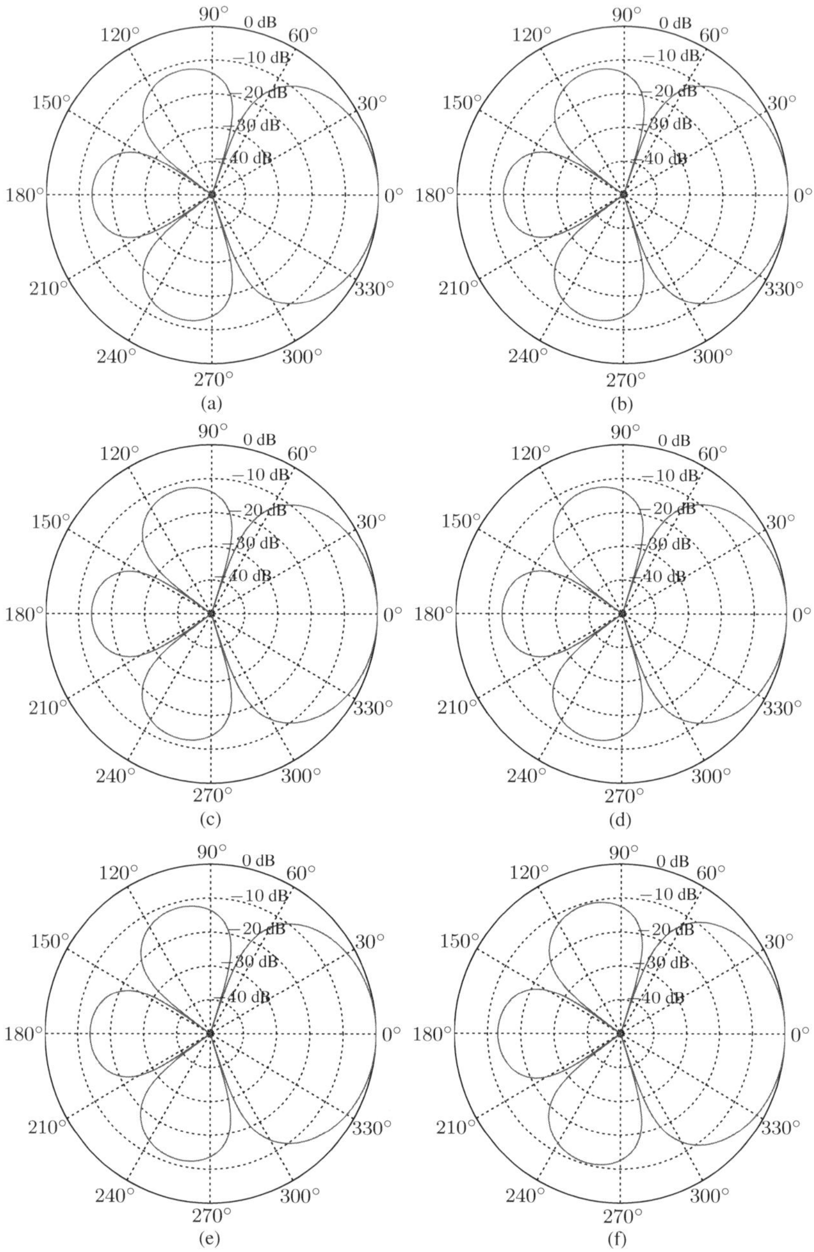

Fig. 4.8 Patterns of the second-order hypercardioid for several frequencies and two values of δ: (a) $f = 0.5$ kHz, $\delta = 1$ cm, (b) $f = 0.5$ kHz, $\delta = 2$ cm, (c) $f = 1$ kHz, $\delta = 1$ cm, (d) $f = 1$ kHz, $\delta = 2$ cm, (e) $f = 3$ kHz, $\delta = 1$ cm, and (f) $f = 3$ kHz, $\delta = 2$ cm.

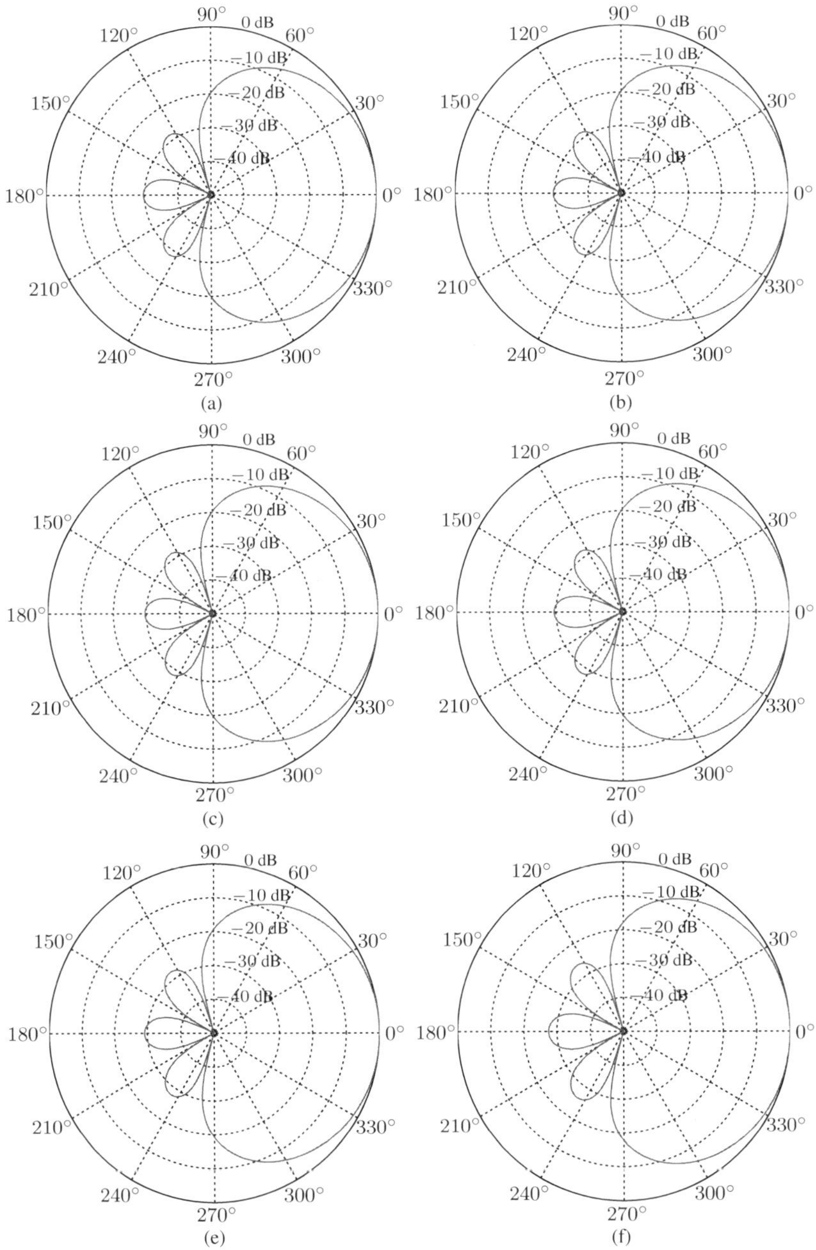

Fig. 4.9 Patterns of the second-order supercardioid for several frequencies and two values of δ: (a) $f = 0.5$ kHz, $\delta = 1$ cm, (b) $f = 0.5$ kHz, $\delta = 2$ cm, (c) $f = 1$ kHz, $\delta = 1$ cm, (d) $f = 1$ kHz, $\delta = 2$ cm, (e) $f = 3$ kHz, $\delta = 1$ cm, and (f) $f = 3$ kHz, $\delta = 2$ cm.

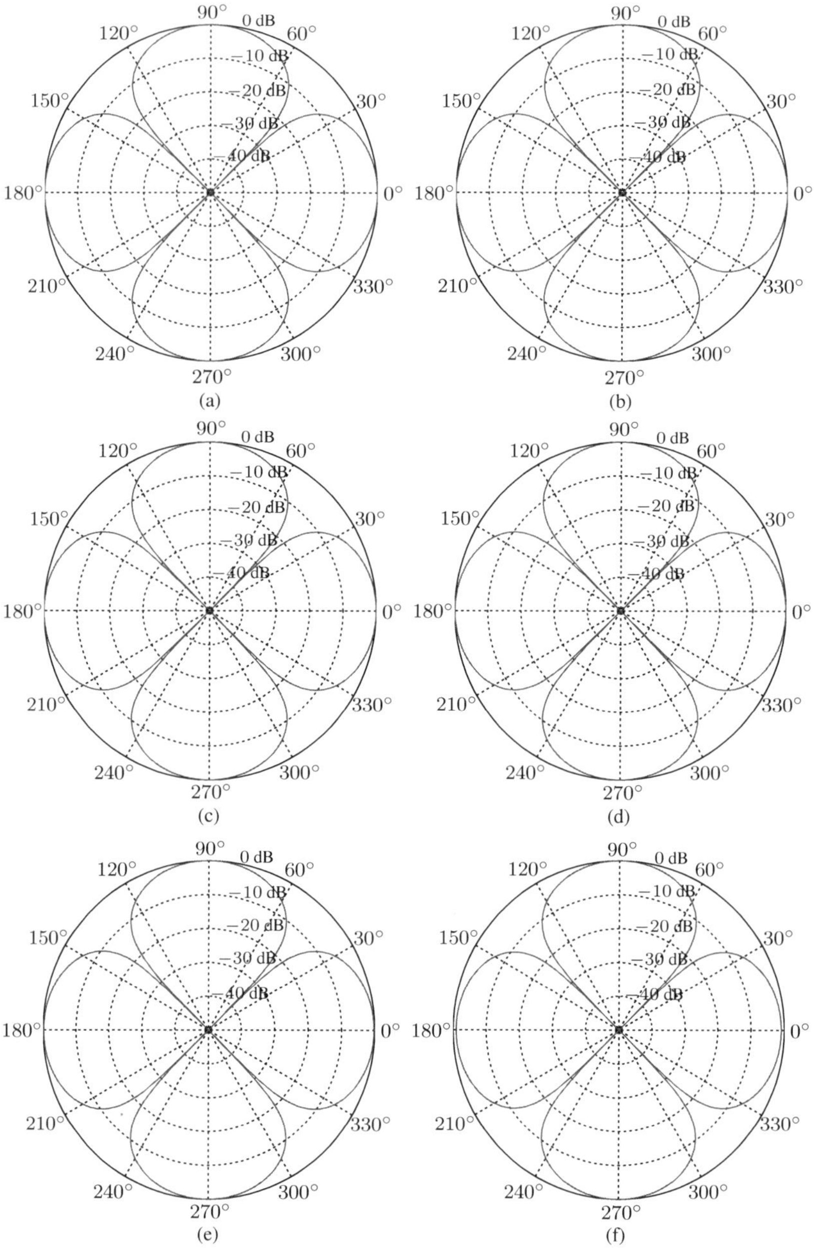

Fig. 4.10 Patterns of the second-order quadrupole for several frequencies and two values of δ: (a) $f = 0.5$ kHz, $\delta = 1$ cm, (b) $f = 0.5$ kHz, $\delta = 2$ cm, (c) $f = 1$ kHz, $\delta = 1$ cm, (d) $f = 1$ kHz, $\delta = 2$ cm, (e) $f = 3$ kHz, $\delta = 1$ cm, and (f) $f = 3$ kHz, $\delta = 2$ cm.

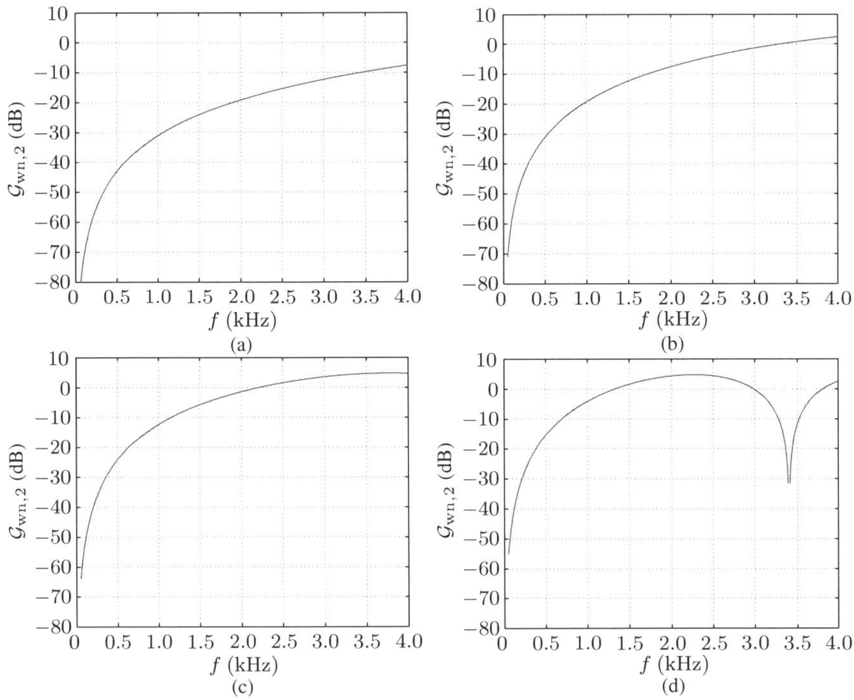

Fig. 4.11 The white noise gain of the second-order cardioid, as a function of frequency, for different values of δ: (a) $\delta = 1$ cm, (b) $\delta = 2$ cm, (c) $\delta = 3$ cm, and (d) $\delta = 5$ cm.

In particular, it can be verified from (4.27) that for the quadrupole, $\mathcal{B}^2\left[\mathbf{h}'(\omega), \theta\right]$ has three identical maximums at the angles 0°, 90°, and 180°.

The white noise gain is

$$\begin{aligned}\mathcal{G}_{\text{wn},2}\left[\mathbf{h}'(\omega)\right] &= \frac{\left|1 - e^{\jmath\omega\tau_0(1-\alpha_{2,1})}\right|^2 \left|1 - e^{\jmath\omega\tau_0(1-\alpha_{2,2})}\right|^2}{4 + 2\cos\left[\omega\tau_0\left(\alpha_{2,1} - \alpha_{2,2}\right)\right]} \\ &= \frac{\left\{1 - \cos\left[\omega\tau_0\left(1 - \alpha_{2,1}\right)\right]\right\}\left\{1 - \cos\left[\omega\tau_0\left(1 - \alpha_{2,2}\right)\right]\right\}}{1 + \frac{1}{2}\cos\left[\omega\tau_0\left(\alpha_{2,1} - \alpha_{2,2}\right)\right]}.\end{aligned} \tag{4.28}$$

In Figs. 4.11, 4.12, 4.13, and 4.14, we plot $\mathcal{G}_{\text{wn},2}\left[\mathbf{h}'(\omega)\right]$ from (4.28), for the cardioid, hypercardioid, supercardioid, and quadrupole, respectively, as a function of frequency, for different values of δ. The white noise gain in (4.28) can be approximated as

$$\mathcal{G}_{\text{wn},2}\left[\mathbf{h}'(\omega)\right] \approx \frac{\left(\omega\tau_0\right)^4\left(1-\alpha_{2,1}\right)^2\left(1-\alpha_{2,2}\right)^2}{6}. \tag{4.29}$$

We deduce that the white noise is amplified if

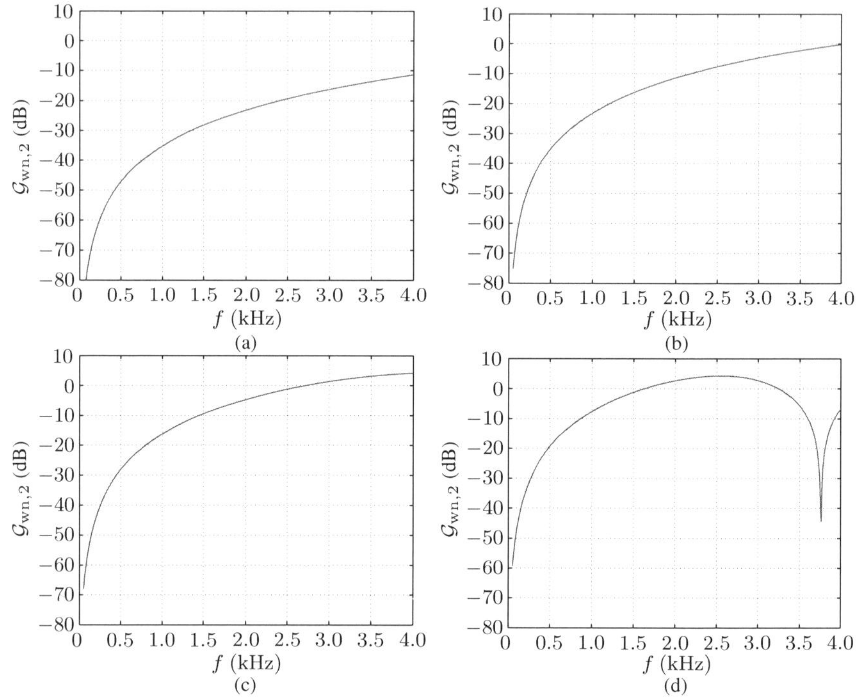

Fig. 4.12 The white noise gain of the second-order hypercardioid, as a function of frequency, for different values of δ: (a) $\delta = 1$ cm, (b) $\delta = 2$ cm, (c) $\delta = 3$ cm, and (d) $\delta = 5$ cm.

$$\omega\tau_0 < \frac{6^{1/4}}{\sqrt{(1-\alpha_{2,1})(1-\alpha_{2,2})}}. \tag{4.30}$$

The directivity factor of the second-order DMA is

$$\mathcal{G}_{\mathrm{dn},2}\left[\mathbf{h}'(\omega)\right] = \frac{\left|1-e^{\jmath\omega\tau_0(1-\alpha_{2,1})}\right|^2 \left|1-e^{\jmath\omega\tau_0(1-\alpha_{2,2})}\right|^2}{D(\omega)}, \tag{4.31}$$

where

$$\begin{aligned} D(\omega) = {} & 4 - 4\,\mathrm{sinc}(\omega\tau_0)\cos(\omega\tau_0\alpha_{2,1}) - 4\,\mathrm{sinc}(\omega\tau_0)\cos(\omega\tau_0\alpha_{2,2}) \\ & + 2\cos\left[\omega\tau_0(\alpha_{2,1}-\alpha_{2,2})\right] \\ & + 2\,\mathrm{sinc}(\omega\tau_0)\cos(\omega\tau_0)\cos\left[\omega\tau_0(\alpha_{2,1}+\alpha_{2,2})\right]. \end{aligned} \tag{4.32}$$

Using Taylor's series, it can be shown that

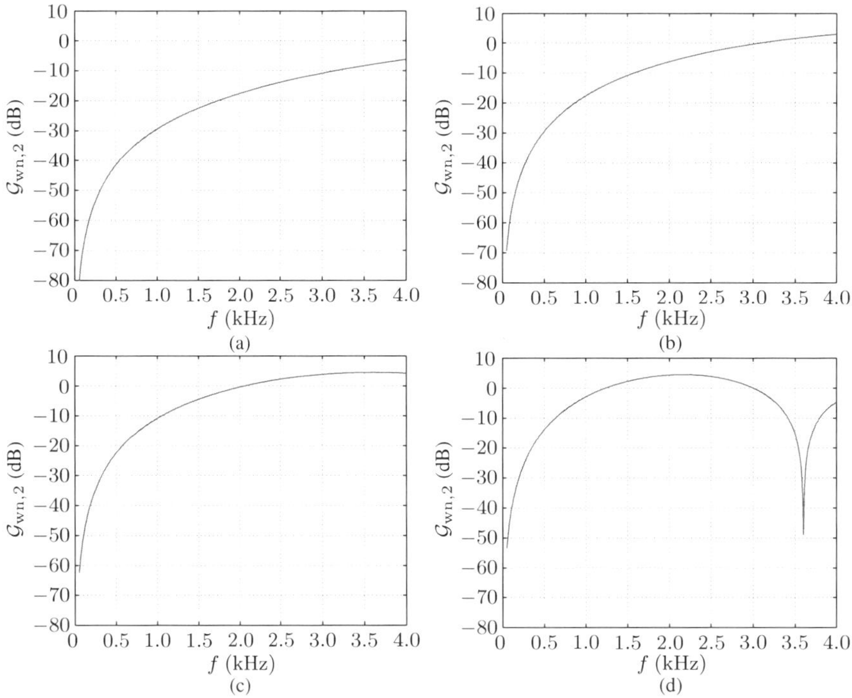

Fig. 4.13 The white noise gain of the second-order supercardioid, as a function of frequency, for different values of δ: (a) $\delta = 1$ cm, (b) $\delta = 2$ cm, (c) $\delta = 3$ cm, and (d) $\delta = 5$ cm.

$$\mathcal{G}_{\mathrm{dn},2}\left[\mathbf{h}'(\omega)\right] \approx 3 \frac{\left(1-\alpha_{2,1}\right)^2\left(1-\alpha_{2,2}\right)^2}{\alpha_{2,1}^2+\alpha_{2,2}^2+4\alpha_{2,1}\alpha_{2,2}+\frac{3}{5}+\frac{1}{2}\alpha_{2,1}^4+\frac{1}{2}\alpha_{2,2}^4+3\alpha_{2,1}^2\alpha_{2,2}^2}. \tag{4.33}$$

Figure 4.15 shows a three dimensional plot of $\mathcal{G}_{\mathrm{dn},2}\left[\mathbf{h}'(\omega)\right]$ from (4.33). We deduce that the theoretical values of the directivity factor for the second-order DMAs studied in this section are as follows.

- Cardioid: $\mathcal{G}_{\mathrm{dn},2}\left[\mathbf{h}'(\omega)\right] \approx 5.7$.
- Hypercardioid: $\mathcal{G}_{\mathrm{dn},2}\left[\mathbf{h}'(\omega)\right] \approx 6.2$.
- Supercardioid: $\mathcal{G}_{\mathrm{dn},2}\left[\mathbf{h}'(\omega)\right] \approx 5.9$.
- Quadrupole: $\mathcal{G}_{\mathrm{dn},2}\left[\mathbf{h}'(\omega)\right] \approx 1.25$.

In Figs. 4.16, 4.17, 4.18, and 4.19, we plot $\mathcal{G}_{\mathrm{dn},2}\left[\mathbf{h}'(\omega)\right]$ from (4.31), for the cardioid, hypercardioid, supercardioid, and quadrupole, respectively, as a function of frequency, for different values of δ.

For a point noise source, we find that the gain is

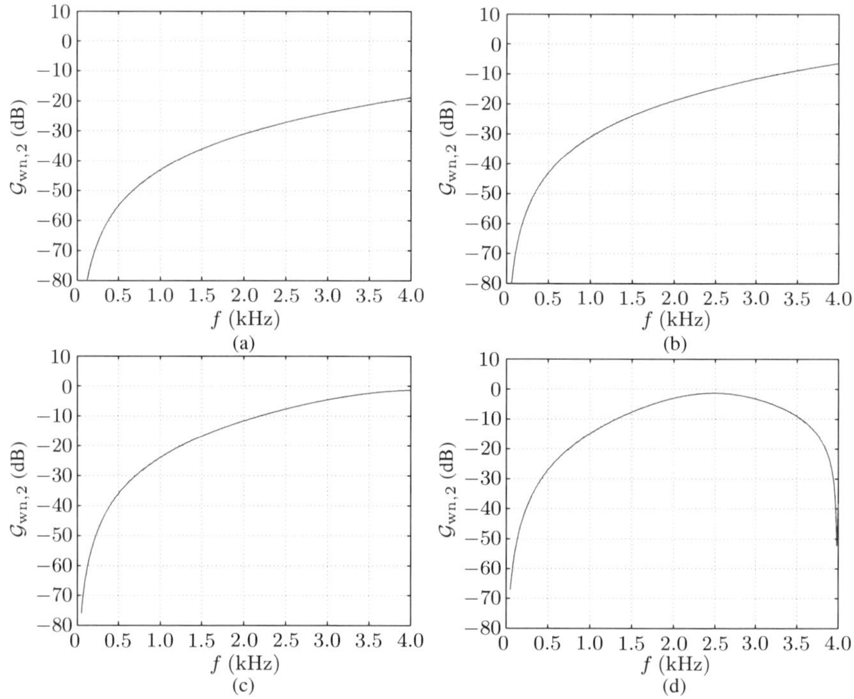

Fig. 4.14 The white noise gain of the second-order quadrupole, as a function of frequency, for different values of δ: (a) $\delta = 1$ cm, (b) $\delta = 2$ cm, (c) $\delta = 3$ cm, and (d) $\delta = 5$ cm.

$$\mathcal{G}_{\mathrm{ns},2}\left[\mathbf{h}'(\omega)\right] = \frac{\left\{1 - \cos\left[\omega\tau_0\left(1 - \alpha_{2,1}\right)\right]\right\}\left\{1 - \cos\left[\omega\tau_0\left(1 - \alpha_{2,2}\right)\right]\right\}}{\left\{1 - \cos\left[\omega\tau_0\left(\cos\theta_{\mathrm{n}} - \alpha_{2,1}\right)\right]\right\}\left\{1 - \cos\left[\omega\tau_0\left(\cos\theta_{\mathrm{n}} - \alpha_{2,2}\right)\right]\right\}}. \tag{4.34}$$

Figures 4.20, 4.21, 4.22, and 4.23 display $\mathcal{G}_{\mathrm{ns},2}\left[\mathbf{h}'(\omega)\right]$, for the cardioid, hypercardioid, supercardioid, and quadrupole, respectively, as a function of θ_{n}, for several frequencies and two values of δ.

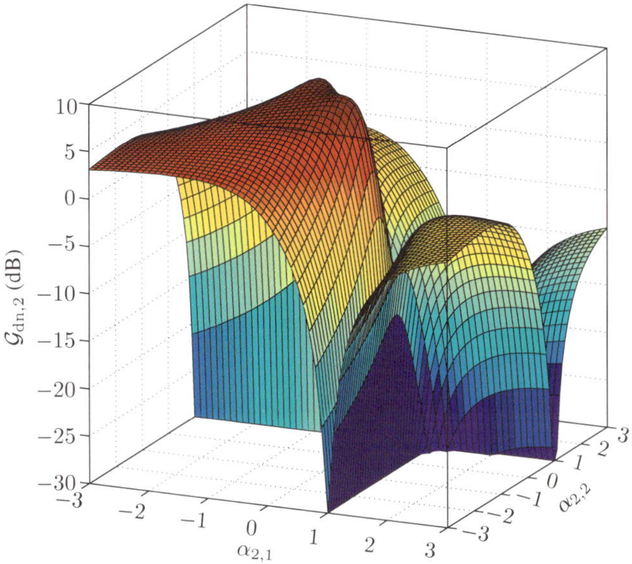

Fig. 4.15 The theoretical directivity factor of the second-order DMA as a function of $\alpha_{2,1}$ and $\alpha_{2,2}$.

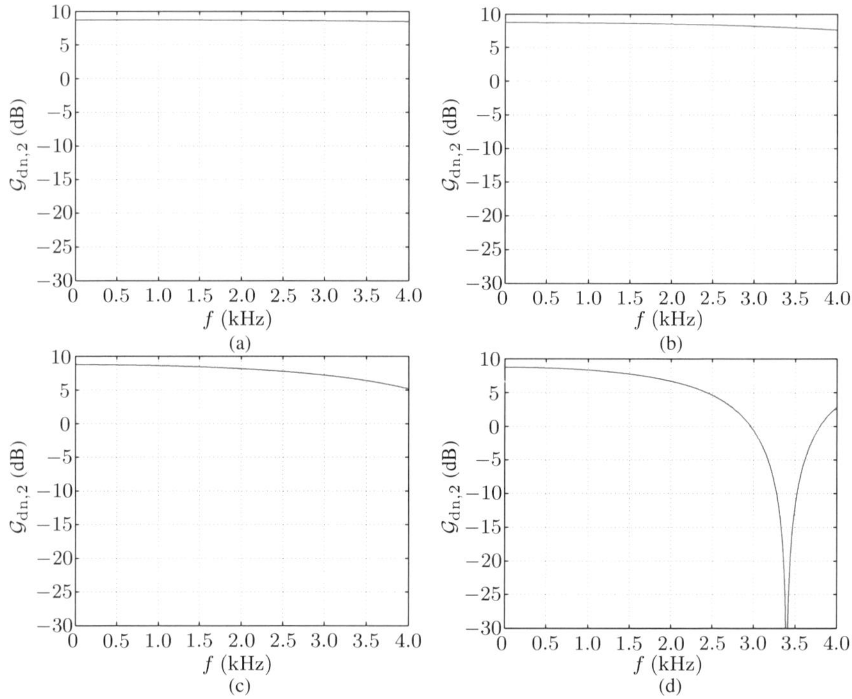

Fig. 4.16 The directivity factor of the second-order cardioid, as a function of frequency, for different values of δ: (a) $\delta = 1$ cm, (b) $\delta = 2$ cm, (c) $\delta = 3$ cm, and (d) $\delta = 5$ cm.

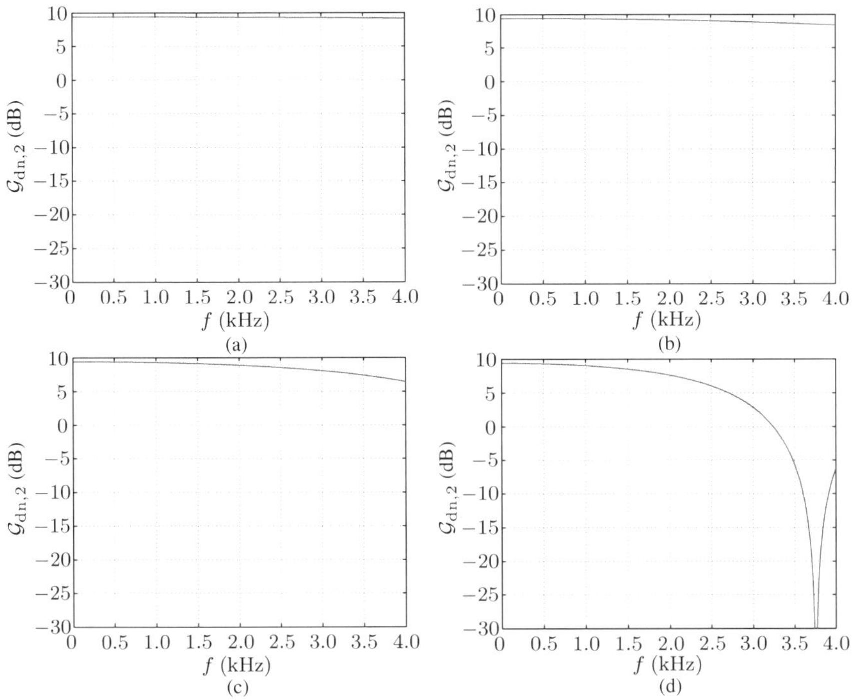

Fig. 4.17 The directivity factor of the second-order hypercardioid, as a function of frequency, for different values of δ: (a) $\delta = 1$ cm, (b) $\delta = 2$ cm, (c) $\delta = 3$ cm, and (d) $\delta = 5$ cm.

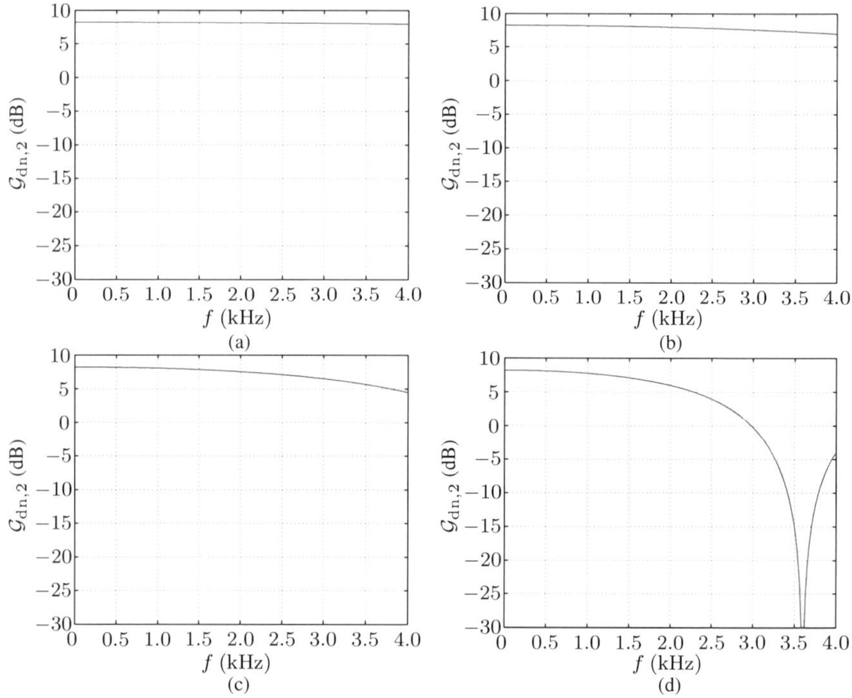

Fig. 4.18 The directivity factor of the second-order supercardioid, as a function of frequency, for different values of δ: (a) $\delta = 1$ cm, (b) $\delta = 2$ cm, (c) $\delta = 3$ cm, and (d) $\delta = 5$ cm.

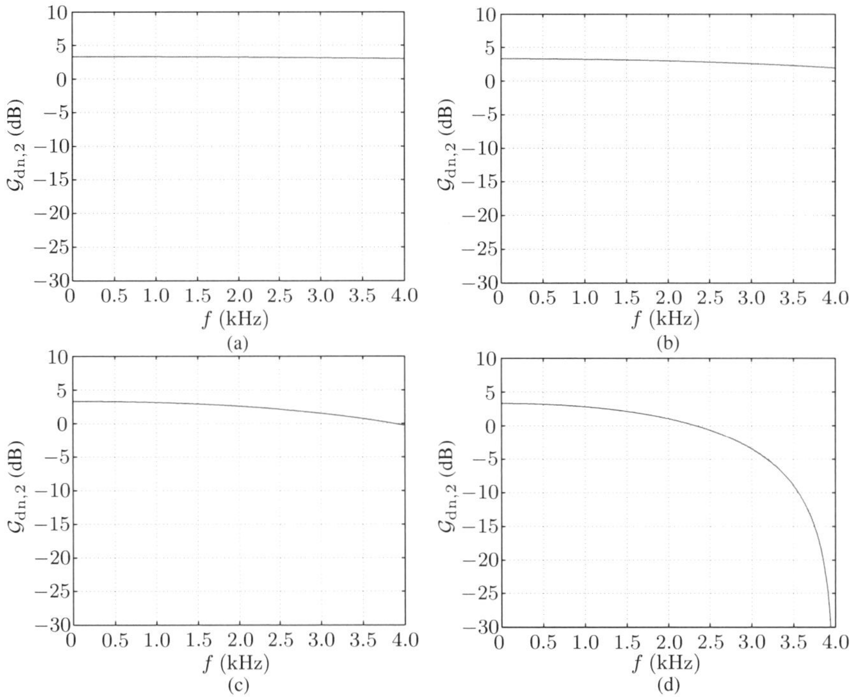

Fig. 4.19 The directivity factor of the second-order quadrupole, as a function of frequency, for different values of δ: (a) $\delta = 1$ cm, (b) $\delta = 2$ cm, (c) $\delta = 3$ cm, and (d) $\delta = 5$ cm.

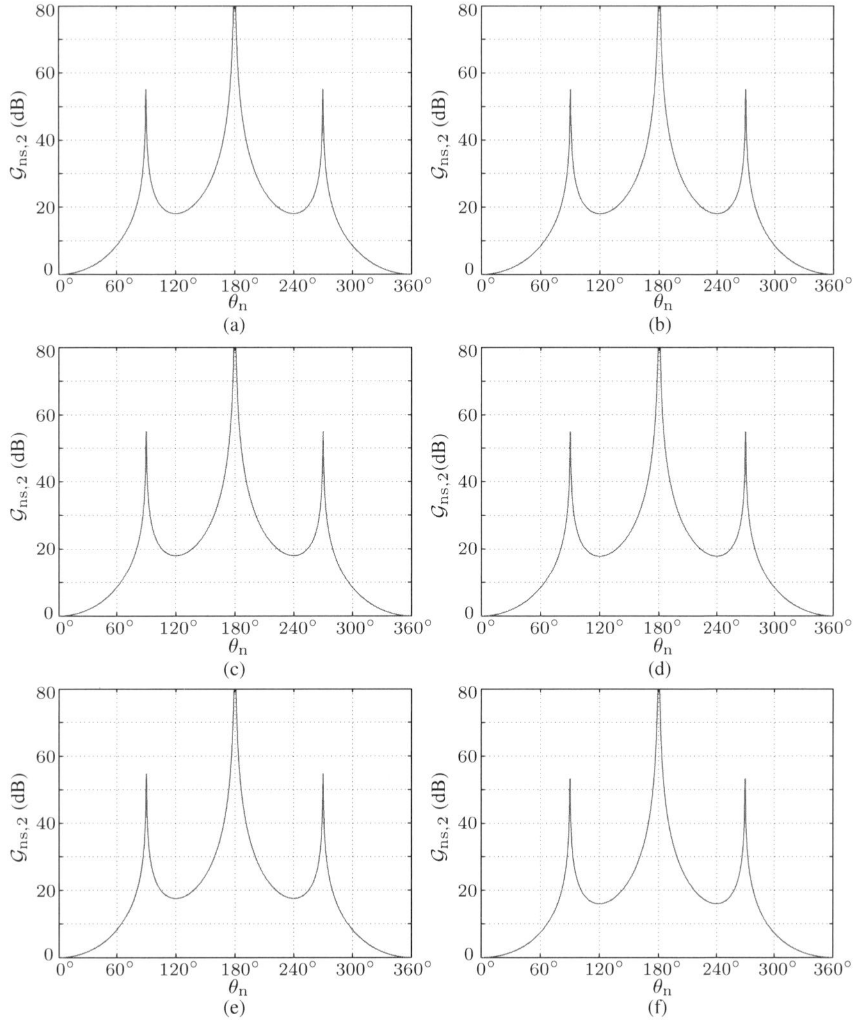

Fig. 4.20 The gain of the second-order cardioid for a point noise source, as a function of θ_{n}, for several frequencies and two values of δ: (a) $f = 0.5$ kHz, $\delta = 1$ cm, (b) $f = 0.5$ kHz, $\delta = 2$ cm, (c) $f = 1$ kHz, $\delta = 1$ cm, (d) $f = 1$ kHz, $\delta = 2$ cm, (e) $f = 3$ kHz, $\delta = 1$ cm, and (f) $f = 3$ kHz, $\delta = 2$ cm.

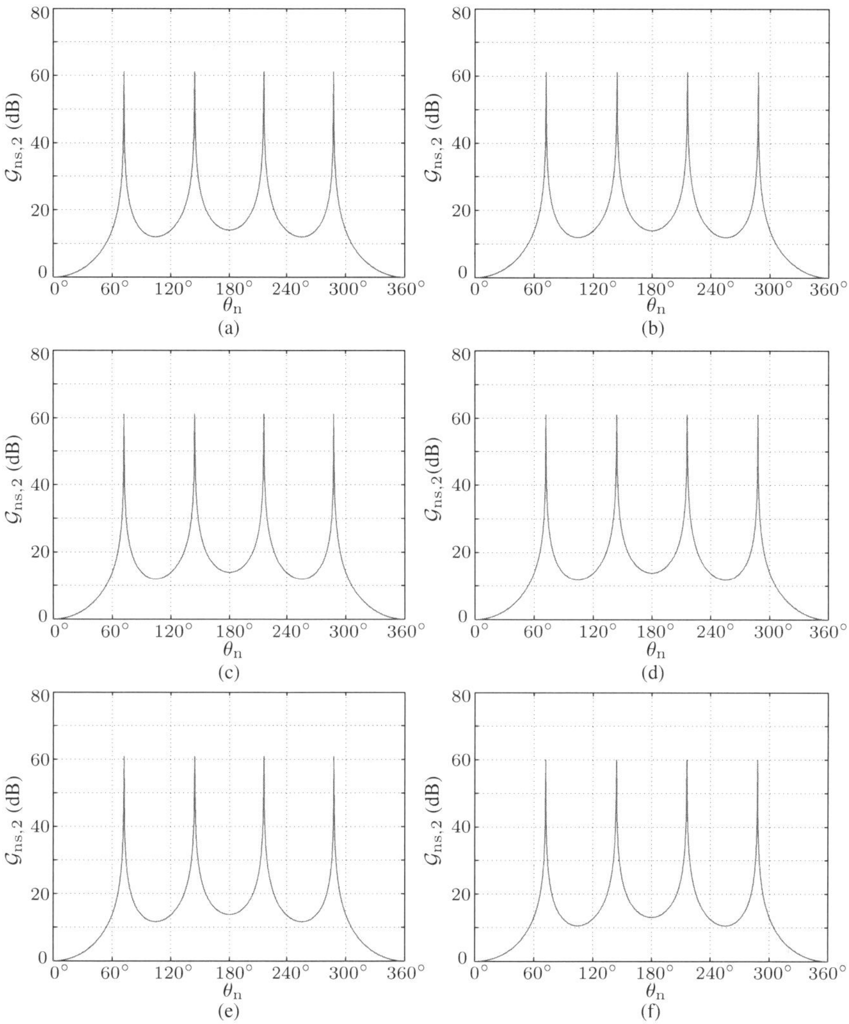

Fig. 4.21 The gain of the second-order hypercardioid for a point noise source, as a function of θ_n, for several frequencies and two values of δ: (a) $f = 0.5$ kHz, $\delta = 1$ cm, (b) $f = 0.5$ kHz, $\delta = 2$ cm, (c) $f = 1$ kHz, $\delta = 1$ cm, (d) $f = 1$ kHz, $\delta = 2$ cm, (e) $f = 3$ kHz, $\delta = 1$ cm, and (f) $f = 3$ kHz, $\delta = 2$ cm.

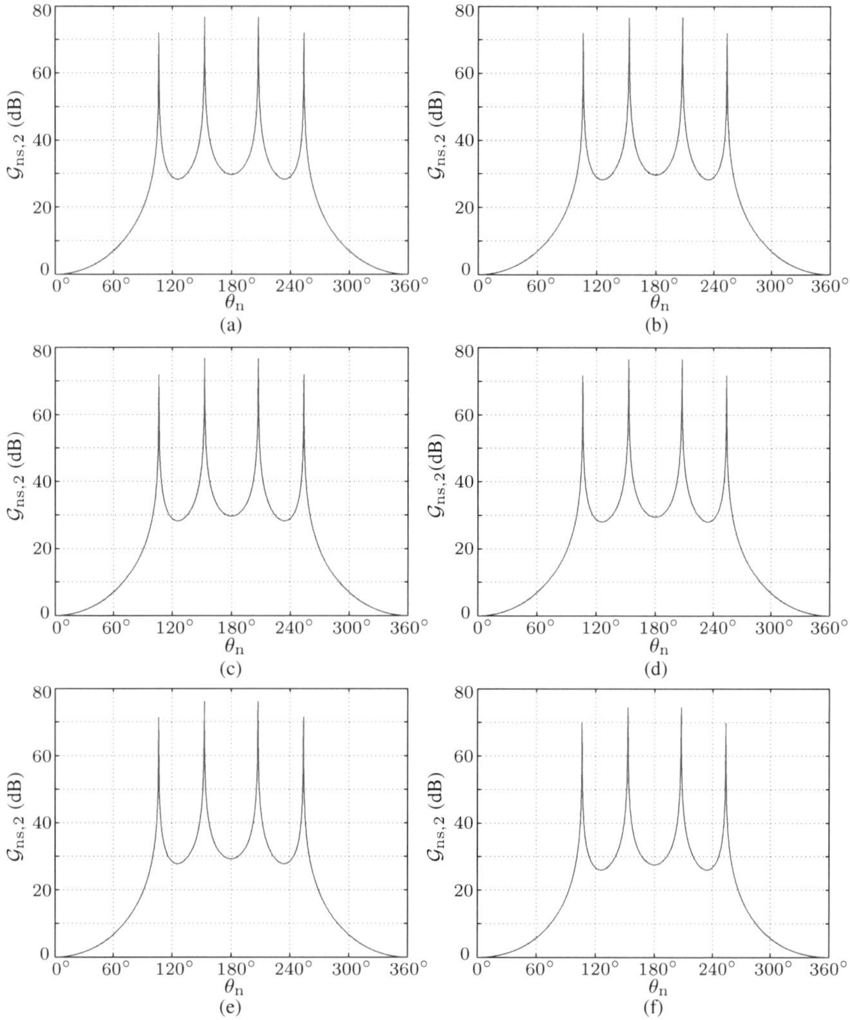

Fig. 4.22 The gain of the second-order supercardioid for a point noise source, as a function of θ_{n}, for several frequencies and two values of δ: (a) $f = 0.5$ kHz, $\delta = 1$ cm, (b) $f = 0.5$ kHz, $\delta = 2$ cm, (c) $f = 1$ kHz, $\delta = 1$ cm, (d) $f = 1$ kHz, $\delta = 2$ cm, (e) $f = 3$ kHz, $\delta = 1$ cm, and (f) $f = 3$ kHz, $\delta = 2$ cm.

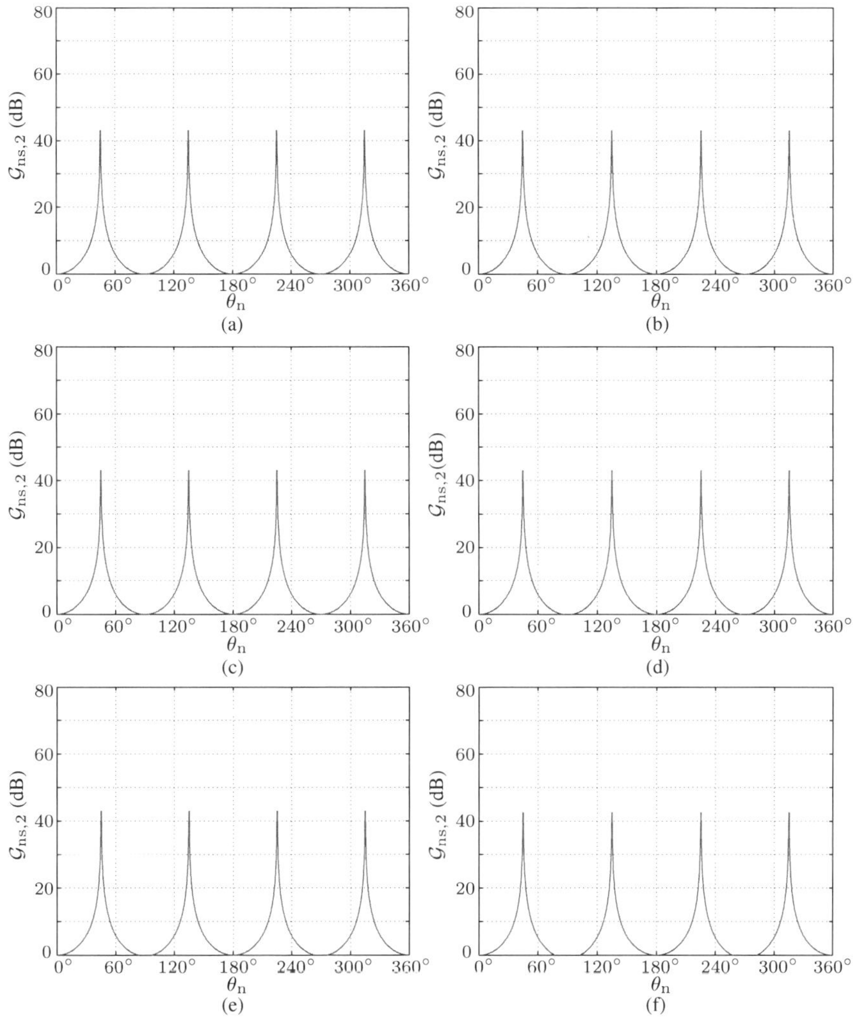

Fig. 4.23 The gain of the second-order quadrupole for a point noise source, as a function of θ_{n}, for several frequencies and two values of δ: (a) $f = 0.5$ kHz, $\delta = 1$ cm, (b) $f = 0.5$ kHz, $\delta = 2$ cm, (c) $f = 1$ kHz, $\delta = 1$ cm, (d) $f = 1$ kHz, $\delta = 2$ cm, (e) $f = 3$ kHz, $\delta = 1$ cm, and (f) $f = 3$ kHz, $\delta = 2$ cm.

4.4 Design of Another Well-Known Second-Order Cardioid

There exists another very well-known second-order cardioid that we have not discussed so far. Its directional pattern is [1], [2]

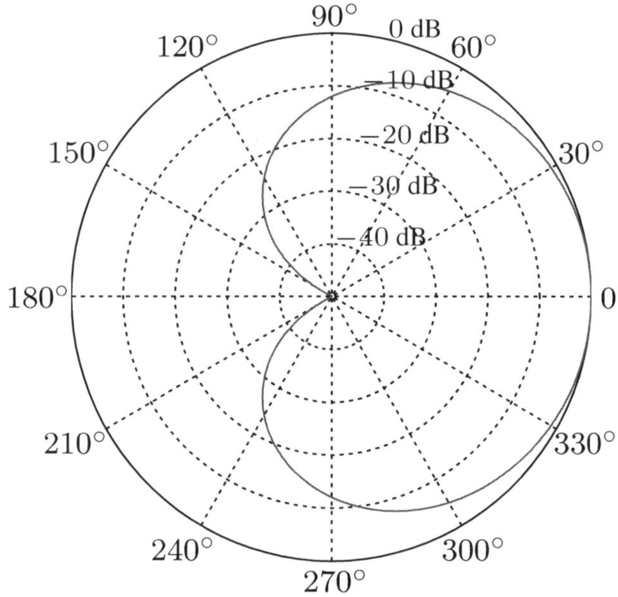

Fig. 4.24 Directional pattern of the second-order cardioid with one distinct null.

$$\mathcal{B}_{\mathrm{C},2}(\theta) = \left(\frac{1}{2} + \frac{1}{2}\cos\theta\right)^2. \tag{4.35}$$

This cardioid has only one null at 180°, while the one studied in the previous section has two distinct nulls (at 90° and 180°). Figure 4.24 shows a plot of $\mathcal{B}^2_{\mathrm{C},2}(\theta)$ in dB.

We can use our approach to design such a DMA by choosing three constraints. The first two are a one at 0° and a null at 180°. From a theoretical point of view, any value on the pattern can play the role of the third constraint but because of the approximations involved, this value must be chosen with care in order that the derived filter corresponds to the original design. Here we take the angle[3] 60° ($\cos 60° = 1/2$), so that $\mathcal{B}_{\mathrm{C},2}(60°) = 9/16$. Therefore, the linear system to solve is

$$\begin{bmatrix} \mathbf{d}^H(\omega, 1) \\ \mathbf{d}^H\left(\omega, \dfrac{1}{2}\right) \\ \mathbf{d}^H(\omega, -1) \end{bmatrix} \mathbf{h}(\omega) = \begin{bmatrix} 1 \\ \dfrac{9}{16} \\ 0 \end{bmatrix}. \tag{4.36}$$

[3] Even with this value, we do not get exactly what we are looking for but the obtained directional pattern is also very interesting. This shows, again, the flexibility of the approach. In fact, taking the constraint above 90°, will lead exactly to the desired pattern. We leave this study to the reader.

The solution to (4.36) is

$$\mathbf{h}(\omega) = \frac{1}{\left(e^{\jmath\omega\tau_0/2} - e^{\jmath\omega\tau_0}\right)\left(e^{-\jmath\omega\tau_0/2} - e^{\jmath 3\omega\tau_0/2}\right)} \begin{bmatrix} 1 \\ -e^{-\jmath\omega\tau_0/2} - e^{\jmath\omega\tau_0} \\ e^{\jmath\omega\tau_0/2} \end{bmatrix}$$
$$+ \frac{9}{16\left(1 - e^{\jmath\omega\tau_0}\right)\left(1 - e^{-\jmath\omega\tau_0}\right)} \begin{bmatrix} 1 \\ -e^{-\jmath\omega\tau_0} - e^{\jmath\omega\tau_0} \\ 1 \end{bmatrix}. \quad (4.37)$$

Using the approximation $e^x \approx 1 + x$ in the two factor terms of (4.37), we obtain

$$\mathbf{h}(\omega) \approx -\frac{7}{16\omega^2\tau_0^2} \begin{bmatrix} 1 \\ -e^{\jmath\omega\tau_0} - \frac{16}{7}e^{-\jmath\omega\tau_0/2} + \frac{9}{7}e^{-\jmath\omega\tau_0} \\ -\frac{9}{7} + \frac{16}{7}e^{\jmath\omega\tau_0/2} \end{bmatrix}, \quad (4.38)$$

from which we deduce the equivalent filter[4]:

$$\mathbf{h}'^{*}(\omega) = \frac{1}{\omega^2} \begin{bmatrix} 1 \\ -e^{-\jmath\omega\tau_0} - \frac{16}{7}e^{\jmath\omega\tau_0/2} + \frac{9}{7}e^{\jmath\omega\tau_0} \\ -\frac{9}{7} + \frac{16}{7}e^{-\jmath\omega\tau_0/2} \end{bmatrix}$$
$$= \frac{1}{\omega}\left\{\frac{1}{\omega}\begin{bmatrix} 1 \\ -\frac{16}{7}e^{\jmath\omega\tau_0/2} + \frac{9}{7}e^{\jmath\omega\tau_0} \\ 0 \end{bmatrix} - \frac{e^{-\jmath\omega\tau_0}}{\omega}\begin{bmatrix} 0 \\ 1 \\ -\frac{16}{7}e^{\jmath\omega\tau_0/2} + \frac{9}{7}e^{\jmath\omega\tau_0} \end{bmatrix}\right\}. \quad (4.39)$$

Figure 4.25 shows how this cardioid can be implemented.

It can be verified that the beampattern of the beamformer $\mathbf{h}'(\omega)$ is

$$\mathcal{B}\left[\mathbf{h}'(\omega), \theta\right] = \frac{1}{\omega^2}\left[1 - e^{\jmath\omega\tau_0(\cos\theta + 1)}\right] \times$$
$$\left[1 - \left(\frac{16}{7}e^{-\jmath\omega\tau_0/2} - \frac{9}{7}e^{-\jmath\omega\tau_0}\right)e^{\jmath\omega\tau_0\cos\theta}\right]. \quad (4.40)$$

Figure 4.26 displays the patterns from the previous equation for several frequencies and two values of δ. For small values of $\omega\tau_0$, we can approximate

[4] Notice that this filter is noncausal. Therefore, we need to add a processing delay equal to τ_0.

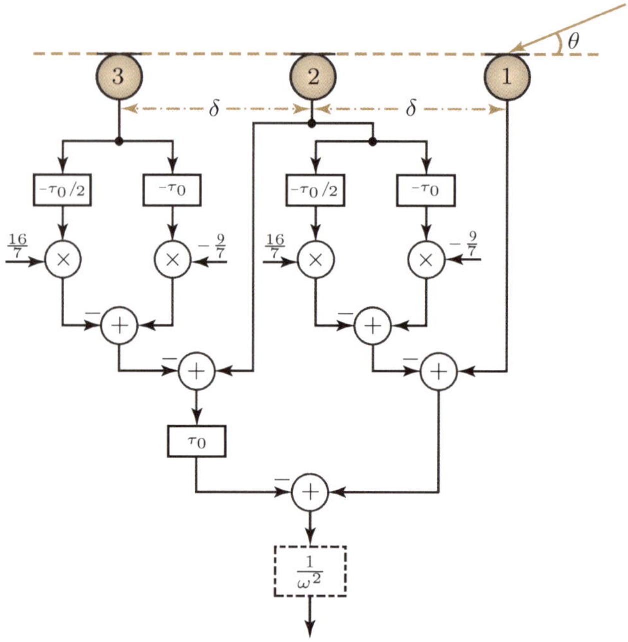

Fig. 4.25 Implementation of the second-order cardioid with one distinct null.

(4.40) as

$$\mathcal{B}\left[\mathbf{h}'(\omega),\theta\right] \approx \frac{4}{\mathcal{C}^2}\left(\frac{1}{2}+\frac{1}{2}\cos\theta\right)^2, \tag{4.41}$$

where we recognize the pattern of the second-order cardioid with one distinct null.

The white noise gain is

$$\mathcal{G}_{\mathrm{wn},2}\left[\mathbf{h}'(\omega)\right] = \frac{\left|1-e^{\jmath 2\omega\tau_0}\right|^2 \left|\frac{16}{7}-\frac{16}{7}e^{\jmath\omega\tau_0/2}\right|^2}{D_{\mathrm{wn},2}(\omega)}, \tag{4.42}$$

where

$$D_{\mathrm{wn},2}(\omega) = 2+\frac{674}{49}-\frac{576}{49}\cos(\omega\tau_0/2)-\frac{18}{7}\cos(2\omega\tau_0)+\frac{32}{7}\cos(3\omega\tau_0/2). \tag{4.43}$$

Figure 4.27 shows plots of $\mathcal{G}_{\mathrm{wn},2}\left[\mathbf{h}'(\omega)\right]$ from (4.42), as a function of frequency, for different values of δ. For small values of $\omega\tau_0$, we get

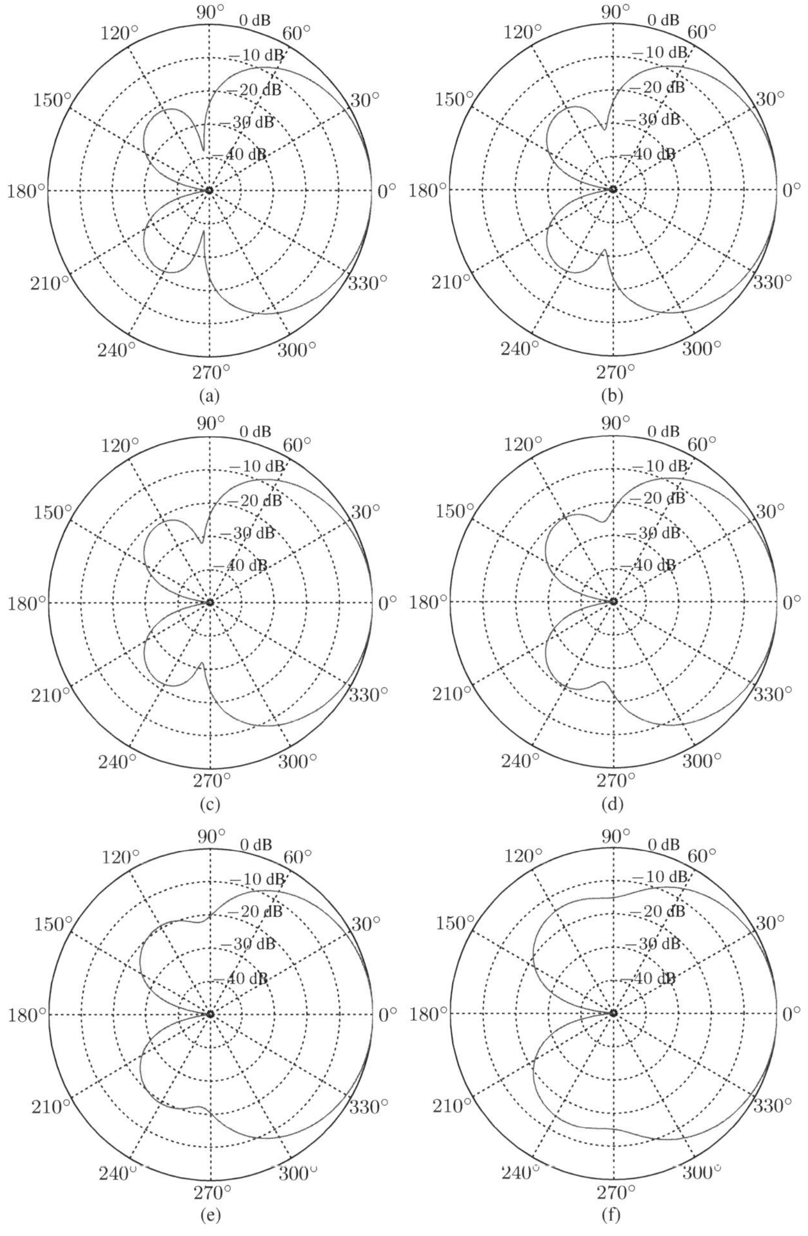

Fig. 4.26 Patterns of the second-order cardioid with one distinct null for several frequencies and two values of δ: (a) $f = 0.5$ kHz, $\delta = 1$ cm, (b) $f = 0.5$ kHz, $\delta = 2$ cm, (c) $f = 1$ kHz, $\delta = 1$ cm, (d) $f = 1$ kHz, $\delta = 2$ cm, (e) $f = 3$ kHz, $\delta = 1$ cm, and (f) $f = 3$ kHz, $\delta = 2$ cm.

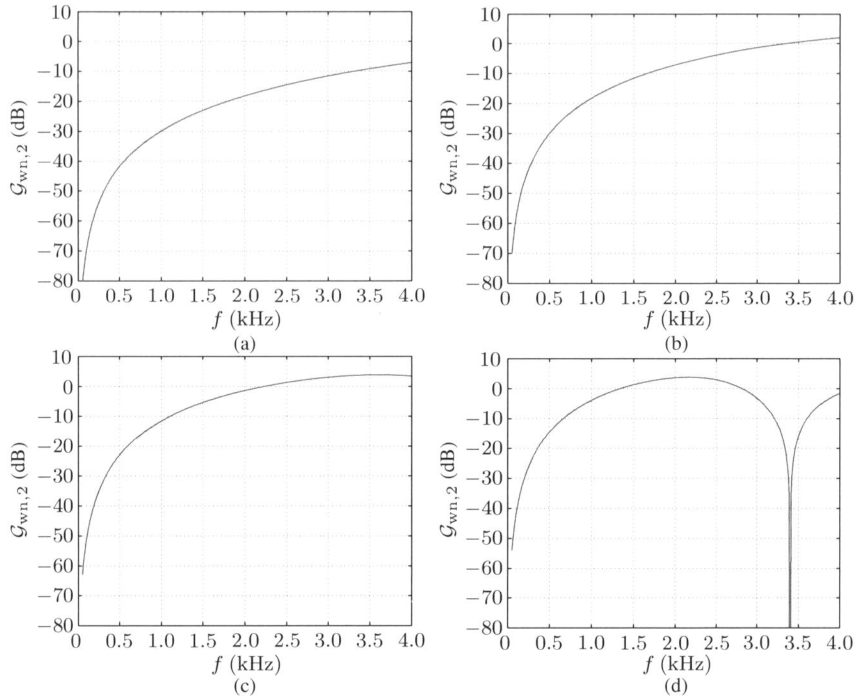

Fig. 4.27 The white noise gain of the second-order cardioid with one distinct null, as a function of frequency, for different values of δ: (a) $\delta = 1$ cm, (b) $\delta = 2$ cm, (c) $\delta = 3$ cm, and (d) $\delta = 5$ cm.

$$\mathcal{G}_{\mathrm{wn},2}\left[\mathbf{h}'(\omega)\right] \approx \frac{256}{294}\left(\omega\tau_0\right)^4, \tag{4.44}$$

from which we deduce that the white noise is amplified if

$$\omega\tau_0 < \left(\frac{294}{256}\right)^{1/4} \approx 1.04. \tag{4.45}$$

The directivity factor of the second-order cardioid with one distinct null is

$$\mathcal{G}_{\mathrm{dn},2}\left[\mathbf{h}'(\omega)\right] = \frac{\left|1 - e^{\jmath 2\omega\tau_0}\right|^2 \left|\frac{16}{7} - \frac{16}{7}e^{\jmath\omega\tau_0/2}\right|^2}{D_{\mathrm{dn},2}(\omega)}, \tag{4.46}$$

where

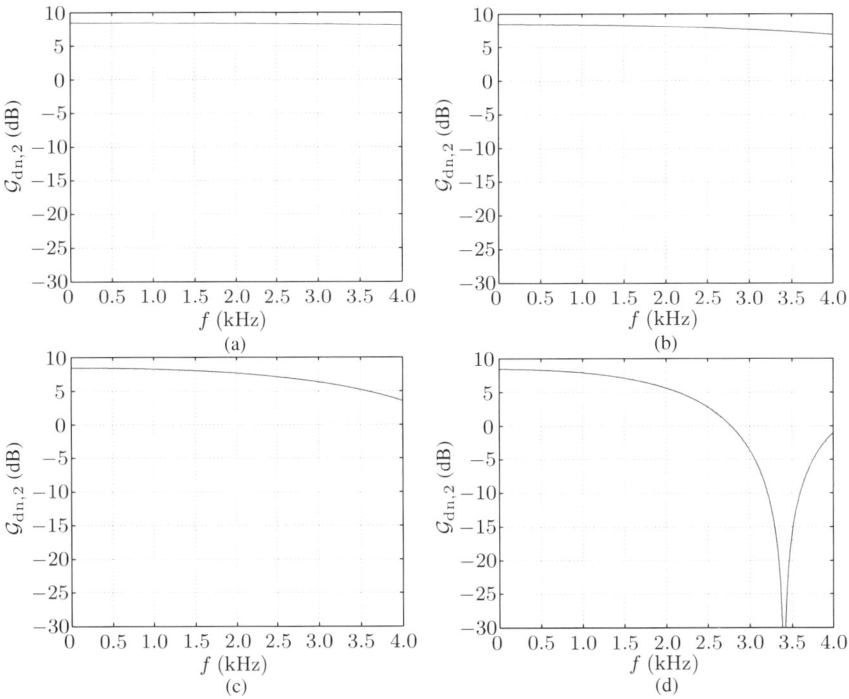

Fig. 4.28 The directivity factor of the second-order cardioid with one distinct null, as a function of frequency, for different values of δ: (a) $\delta = 1$ cm, (b) $\delta = 2$ cm, (c) $\delta = 3$ cm, and (d) $\delta = 5$ cm.

$$\begin{aligned} D_{dn,2}(\omega) = & \frac{772}{49} - \frac{576}{49}\cos(\omega\tau_0/2) - \frac{18}{7}\cos(2\omega\tau_0) + \frac{32}{7}\cos(3\omega\tau_0/2) \\ & + \operatorname{sinc}(\omega\tau_0)\cos(\omega\tau_0)\left[-\frac{18}{7} + \frac{32}{7}\cos(\omega\tau_0/2)\right] \\ & - \operatorname{sinc}(\omega\tau_0)\left[\frac{520}{49}\cos(\omega\tau_0) + \frac{160}{49}\cos(\omega\tau_0/2) - \frac{288}{49}\cos(3\omega\tau_0/2)\right]. \end{aligned} \tag{4.47}$$

In Fig. 4.28, we plot $\mathcal{G}_{dn,2}[\mathbf{h}'(\omega)]$ from (4.46), as a function of frequency, for different values of δ.

Finally, to end this section, we give the gain for a point noise source:

$$\mathcal{G}_{ns,2}[\mathbf{h}'(\omega)] = \frac{\left|1 - e^{\jmath 2\omega\tau_0}\right|^2 \left|\frac{16}{7} - \frac{16}{7}e^{\jmath\omega\tau_0/2}\right|^2}{D_{ns,2}(\omega)}, \tag{4.48}$$

where

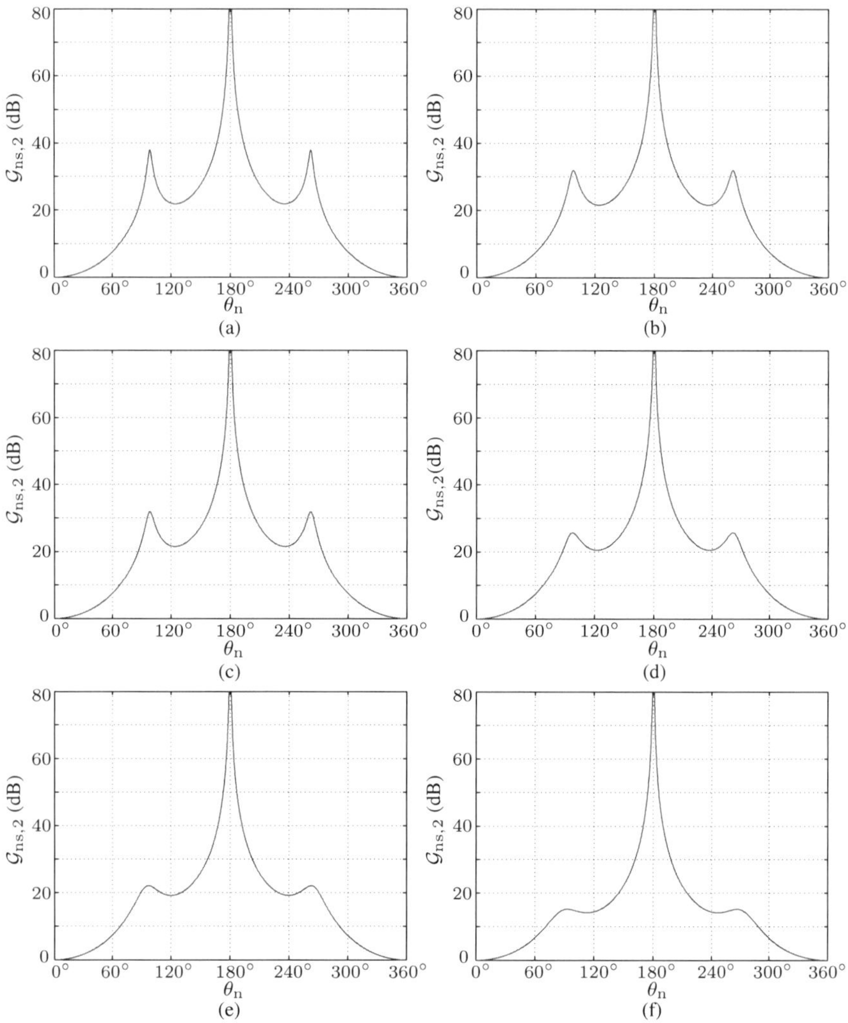

Fig. 4.29 The gain of the second-order cardioid with one distinct null for a point noise source, as a function of θ_{n}, for several frequencies and two values of δ: (a) $f = 0.5$ kHz, $\delta = 1$ cm, (b) $f = 0.5$ kHz, $\delta = 2$ cm, (c) $f = 1$ kHz, $\delta = 1$ cm, (d) $f = 1$ kHz, $\delta = 2$ cm, (e) $f = 3$ kHz, $\delta = 1$ cm, and (f) $f = 3$ kHz, $\delta = 2$ cm.

$$D_{\mathrm{ns},2}(\omega) = \left|1 - e^{\jmath\omega\tau_0(\cos\theta_{\mathrm{n}}+1)}\right|^2 \times \left|1 - \left(\frac{16}{7}e^{-\jmath\omega\tau_0/2} - \frac{9}{7}e^{-\jmath\omega\tau_0}\right)e^{\jmath\omega\tau_0\cos\theta_{\mathrm{n}}}\right|^2. \quad (4.49)$$

Figure 4.29 shows plots of $\mathcal{G}_{\mathrm{ns},2}[\mathbf{h}'(\omega)]$, as a function of θ_{n}, for several frequencies and two values of δ.

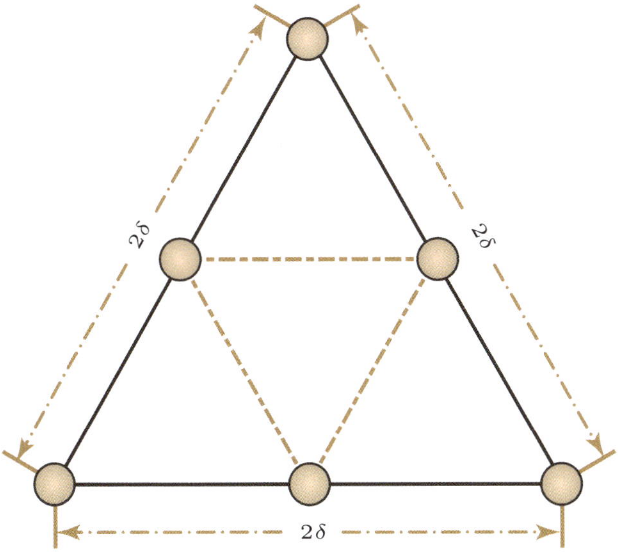

Fig. 4.30 A two-dimensional differential array.

4.5 Steering

As explained in the previous chapter, it is not possible to steer with a uniform linear DMA. One easy way to design any second-order DMA in a certain number of directions, is to double the number of microphones and form an equilateral triangle with three microphones on each one of its sides as shown in Fig. 4.30. Therefore, with this two-dimensional array, we can realize all second-order directional patterns at six different directions: $0°$, $60°$, $120°$, $180°$, $240°$, and $300°$. Clearly, we can also combine other pairs of microphones to realize all first-order directional patterns in many more directions.

References

1. G. W. Elko, "Superdirectional microphone arrays," in *Acoustic Signal Processing for Telecommunication*, S. L. Gay and J. Benesty, Eds. Boston, MA: Kluwer Academic Publishers, 2000, Chapter 10, pp. 181–237.
2. T. D. Abhayapala and A. Gupta, "Higher order differential-integral microphone arrays," *J. Acoust. Soc. Am.*, vol. 127, pp. EL227–EL233, May 2010.

Chapter 5
Study and Design of Third-Order Differential Arrays with Three Distinct Nulls

In this chapter, we study an important family of third-order differential microphone arrays. In this family, which includes the hypercardioid and supercardioid, the directional patterns have exactly three distinct nulls. This feature simplifies the design of third-order DMAs. We will also see how this approach can be easily generalized to the Nth order.

5.1 Principle

Any third-order directional pattern with three distinct nulls can be generated with a uniform linear array of four microphones [1], [2]. It is clear that the linear system of four equations tailored for the derivation of a beamformer with three distinct nulls is

$$\begin{bmatrix} \mathbf{d}^H(\omega, 1) \\ \mathbf{d}^H(\omega, \alpha_{3,1}) \\ \mathbf{d}^H(\omega, \alpha_{3,2}) \\ \mathbf{d}^H(\omega, \alpha_{3,3}) \end{bmatrix} \mathbf{h}(\omega) = \begin{bmatrix} 1 \\ 0 \\ 0 \\ 0 \end{bmatrix}, \tag{5.1}$$

where $-1 \leq \alpha_{3,1} < 1$, $-1 \leq \alpha_{3,2} < 1$, $-1 \leq \alpha_{3,3} < 1$, and $\alpha_{3,1} \neq \alpha_{3,2} \neq \alpha_{3,3}$. We denote by

$$\mathbf{V}_4 = \begin{bmatrix} \mathbf{d}^H(\omega, 1) \\ \mathbf{d}^H(\omega, \alpha_{3,1}) \\ \mathbf{d}^H(\omega, \alpha_{3,2}) \\ \mathbf{d}^H(\omega, \alpha_{3,3}) \end{bmatrix} = \begin{bmatrix} 1 & v_1 & v_1^2 & v_1^3 \\ 1 & v_2 & v_2^2 & v_2^3 \\ 1 & v_3 & v_3^2 & v_3^3 \\ 1 & v_4 & v_4^2 & v_4^3 \end{bmatrix} \tag{5.2}$$

J. Benesty and J. Chen, *Study and Design of Differential Microphone Arrays*,
Springer Topics in Signal Processing 6, DOI: 10.1007/978-3-642-33753-6_5,

the 4×4 Vandermonde matrix that appears in (5.1), where $v_1 = e^{\jmath\omega\tau_0}$, $v_2 = e^{\jmath\omega\tau_0\alpha_{3,1}}$, $v_3 = e^{\jmath\omega\tau_0\alpha_{3,2}}$, and $v_4 = e^{\jmath\omega\tau_0\alpha_{3,3}}$. Because of the structure of the vector on the right-hand side of (5.1), we only need to compute the first column of $\mathbf{V}_4^{-1}$ to find $\mathbf{h}(\omega)$. Using the decomposition $\mathbf{V}_4^{-1} = \mathbf{U}_4\mathbf{L}_4$, the matrix $\mathbf{U}_4$, and the first column of $\mathbf{L}_4$, we find that the first column of $\mathbf{V}_4^{-1}$ is

$$\mathbf{V}_4^{-1}(:,1) = \begin{bmatrix} \dfrac{v_2v_3v_4}{(v_2-v_1)(v_3-v_1)(v_4-v_1)} \\ -\dfrac{v_2v_3+v_3v_4+v_2v_4}{(v_2-v_1)(v_3-v_1)(v_4-v_1)} \\ \dfrac{v_2+v_3+v_4}{(v_2-v_1)(v_3-v_1)(v_4-v_1)} \\ -\dfrac{1}{(v_2-v_1)(v_3-v_1)(v_4-v_1)} \end{bmatrix}. \quad (5.3)$$

We will be interested in three particular cases for the values of the coefficients $\alpha_{3,1}$, $\alpha_{3,2}$, and $\alpha_{3,3}$.

- Case 1: $\alpha_{3,1} = 0$, $\alpha_{3,2} = -\frac{1}{2}$, and $\alpha_{3,3} = -1$.
- Case 2: $\alpha_{3,1} = \frac{1}{2}$, $\alpha_{3,2} = -\frac{1}{2}$, and $\alpha_{3,3} = -1$.
- Case 3: $\alpha_{3,1} = \frac{\sqrt{2}}{2}$, $\alpha_{3,2} = \frac{\sqrt{2}-2}{2}$, and $\alpha_{3,3} = \frac{\sqrt{2}-\sqrt{10}}{2}$.

5.2 Differential Arrays with Three Distinct Nulls

In the context where a one at the angle 0° and three distinct nulls at the angles $\arccos\alpha_{3,1}$, $\arccos\alpha_{3,2}$, and $\arccos\alpha_{3,3}$ are desired, we need to solve the linear system given in (5.1). From $\mathbf{V}_4^{-1}(:,1)$, we easily find that the solution is

$$\mathbf{h}(\omega) = \frac{1}{\left[1-e^{\jmath\omega\tau_0(1-\alpha_{3,1})}\right]\left[1-e^{\jmath\omega\tau_0(1-\alpha_{3,2})}\right]\left[1-e^{\jmath\omega\tau_0(1-\alpha_{3,3})}\right]} \times \begin{bmatrix} 1 \\ -e^{-\jmath\omega\tau_0\alpha_{3,1}} - e^{-\jmath\omega\tau_0\alpha_{3,2}} - e^{-\jmath\omega\tau_0\alpha_{3,3}} \\ e^{-\jmath\omega\tau_0(\alpha_{3,1}+\alpha_{3,2})} + e^{-\jmath\omega\tau_0(\alpha_{3,2}+\alpha_{3,3})} + e^{-\jmath\omega\tau_0(\alpha_{3,1}+\alpha_{3,3})} \\ -e^{-\jmath\omega\tau_0(\alpha_{3,1}+\alpha_{3,2}+\alpha_{3,3})} \end{bmatrix}. \quad (5.4)$$

Using the approximation $e^x \approx 1 + x$ in the factor term of (5.4), we obtain

$$\mathbf{h}(\omega) \approx \frac{\mathcal{C}^3}{\omega^3 (1-\alpha_{3,1})(1-\alpha_{3,2})(1-\alpha_{3,3})} \times \begin{bmatrix} 1 \\ -e^{-\jmath\omega\tau_0\alpha_{3,1}} - e^{-\jmath\omega\tau_0\alpha_{3,2}} - e^{-\jmath\omega\tau_0\alpha_{3,3}} \\ e^{-\jmath\omega\tau_0(\alpha_{3,1}+\alpha_{3,2})} + e^{-\jmath\omega\tau_0(\alpha_{3,2}+\alpha_{3,3})} + e^{-\jmath\omega\tau_0(\alpha_{3,1}+\alpha_{3,3})} \\ -e^{-\jmath\omega\tau_0(\alpha_{3,1}+\alpha_{3,2}+\alpha_{3,3})} \end{bmatrix}, \tag{5.5}$$

where the term $1/\omega^3$ is a third-order low-pass filter. We deduce that the equivalent filter is[1]

$$\mathbf{h}'^{*}(\omega) = \frac{1}{\omega^3} \times \begin{bmatrix} 1 \\ -e^{\jmath\omega\tau_0\alpha_{3,1}} - e^{\jmath\omega\tau_0\alpha_{3,2}} - e^{\jmath\omega\tau_0\alpha_{3,3}} \\ e^{\jmath\omega\tau_0(\alpha_{3,1}+\alpha_{3,2})} + e^{\jmath\omega\tau_0(\alpha_{3,2}+\alpha_{3,3})} + e^{\jmath\omega\tau_0(\alpha_{3,1}+\alpha_{3,3})} \\ -e^{\jmath\omega\tau_0(\alpha_{3,1}+\alpha_{3,2}+\alpha_{3,3})} \end{bmatrix}. \tag{5.6}$$

Figure 5.1 illustrates the implementation of the third-order differential array.

It can be verified that the beampattern of the beamformer $\mathbf{h}'(\omega)$ is

$$\mathcal{B}[\mathbf{h}'(\omega), \theta] = \frac{1}{\omega^3} \prod_{n=1}^{3} \left[1 - e^{\jmath\omega\tau_0(\cos\theta - \alpha_{3,n})}\right], \tag{5.7}$$

which can be approximated as the frequency-independent second-order directional pattern:

$$\mathcal{B}[\mathbf{h}'(\omega), \theta] \approx \frac{1}{\mathcal{C}^3} \prod_{n=1}^{3} (\cos\theta - \alpha_{3,n}). \tag{5.8}$$

Figures 5.2, 5.3, and 5.4 display the patterns from (5.7), of Case 1, Case 2, and Case 3, respectively, for several frequencies and two values of δ.

The white noise gain is

[1] Notice that this filter is noncausal for $\alpha_{3,1} > 0$ or $\alpha_{3,2} > 0$ or $\alpha_{3,3} > 0$. In this situation, a processing delay has to be added.

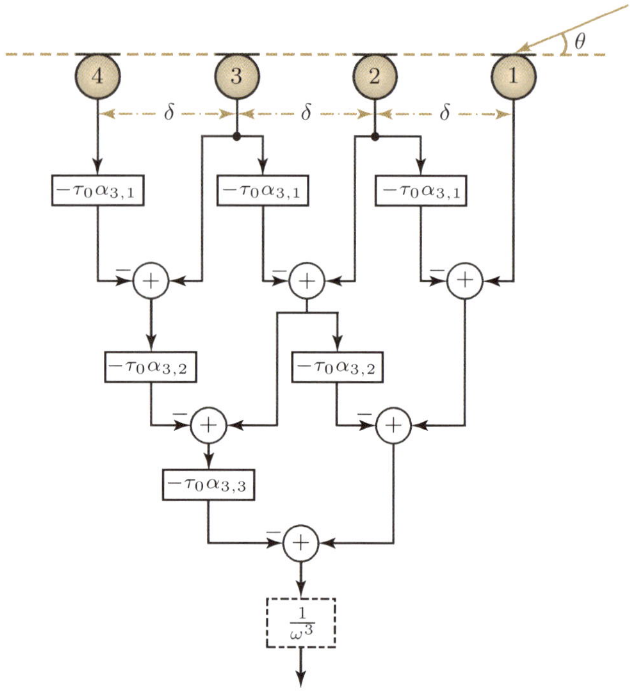

Fig. 5.1 Implementation of the third-order differential array.

$$\mathcal{G}_{\text{wn},3}\left[\mathbf{h}'(\omega)\right] = \frac{\prod_{n=1}^{3}\left|1 - e^{\jmath\omega\tau_0\left(1-\alpha_{3,n}\right)}\right|^2}{8 + 4\sum_i\sum_{j>i}\cos\left[\omega\tau_0\left(\alpha_{3,j}-\alpha_{3,i}\right)\right]} \tag{5.9}$$
$$= \frac{\prod_{n=1}^{3}\left\{1 - \cos\left[\omega\tau_0\left(1-\alpha_{3,n}\right)\right]\right\}}{1 + \frac{1}{2}\sum_i\sum_{j>i}\cos\left[\omega\tau_0\left(\alpha_{3,j}-\alpha_{3,i}\right)\right]}.$$

In Figs 5.5, 5.6, and 5.7, we plot the white noise gain from the previous expression, of Case 1, Case 2, and Case 3, respectively, as a function of frequency, for different values of δ. We can approximate (5.9) as

$$\mathcal{G}_{\text{wn},3}\left[\mathbf{h}'(\omega)\right] \approx \frac{\left(\omega\tau_0\right)^6\prod_{n=1}^{3}\left(1-\alpha_{3,n}\right)^2}{12}. \tag{5.10}$$

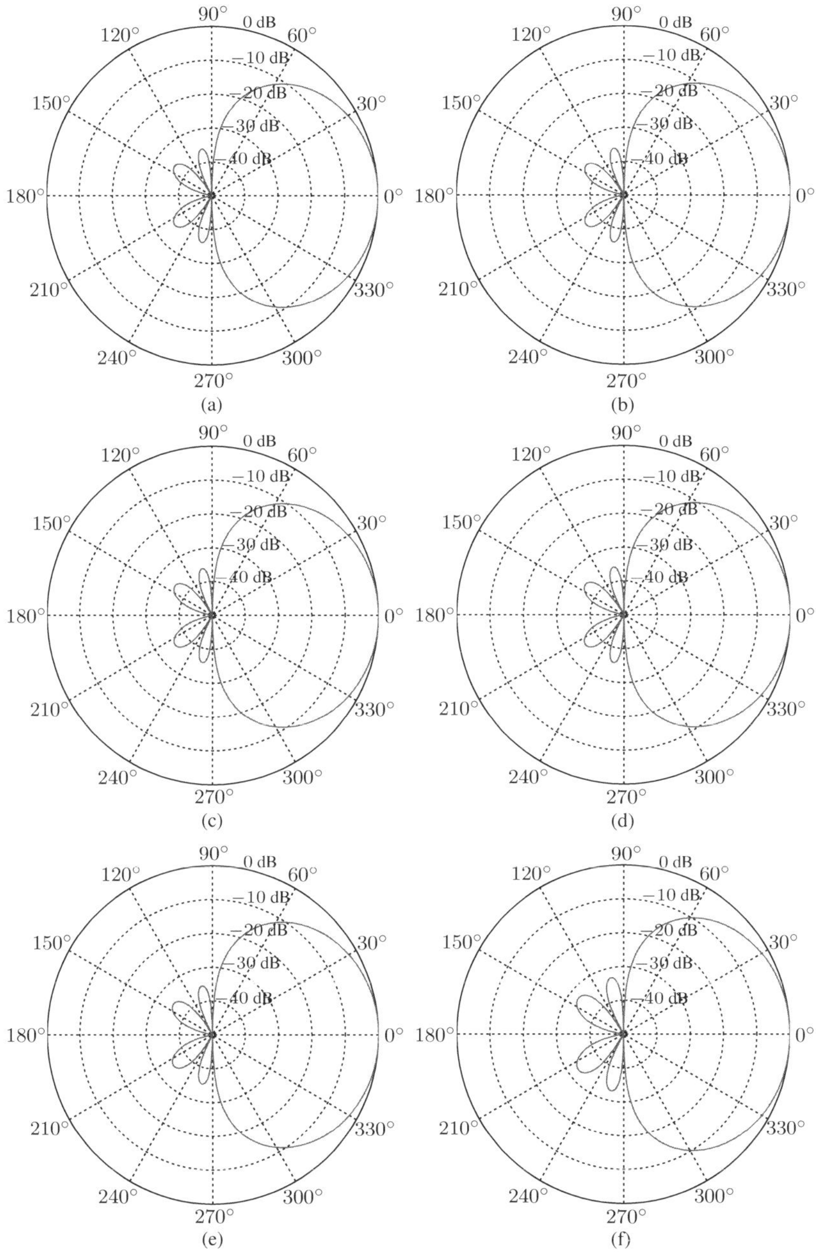

Fig. 5.2 Patterns of Case 1 for several frequencies and two values of δ: (a) $f = 0.5$ kHz, $\delta = 1$ cm, (b) $f = 0.5$ kHz, $\delta = 2$ cm, (c) $f = 1$ kHz, $\delta = 1$ cm, (d) $f = 1$ kHz, $\delta = 2$ cm, (e) $f = 3$ kHz, $\delta = 1$ cm, and (f) $f = 3$ kHz, $\delta = 2$ cm.

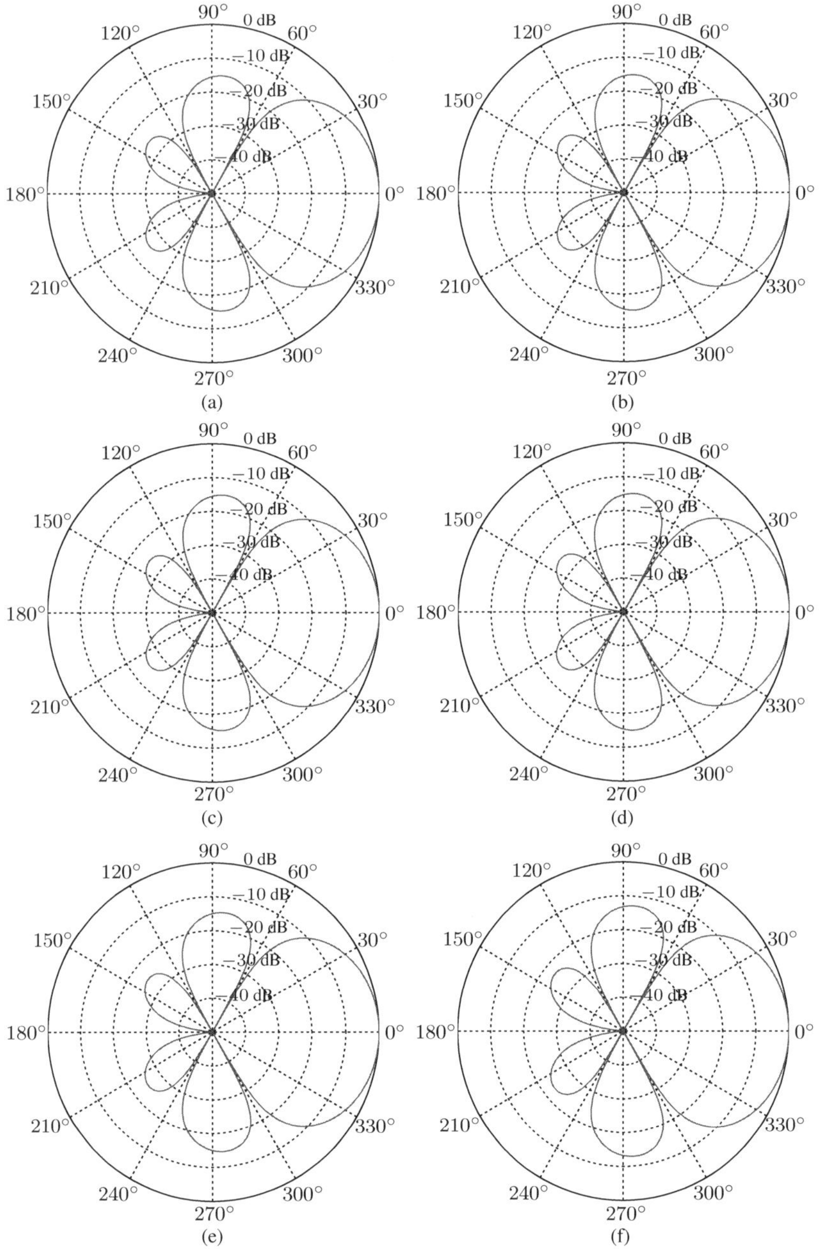

Fig. 5.3 Patterns of Case 2 for several frequencies and two values of δ: (a) $f = 0.5$ kHz, $\delta = 1$ cm, (b) $f = 0.5$ kHz, $\delta = 2$ cm, (c) $f = 1$ kHz, $\delta = 1$ cm, (d) $f = 1$ kHz, $\delta = 2$ cm, (e) $f = 3$ kHz, $\delta = 1$ cm, and (f) $f = 3$ kHz, $\delta = 2$ cm.

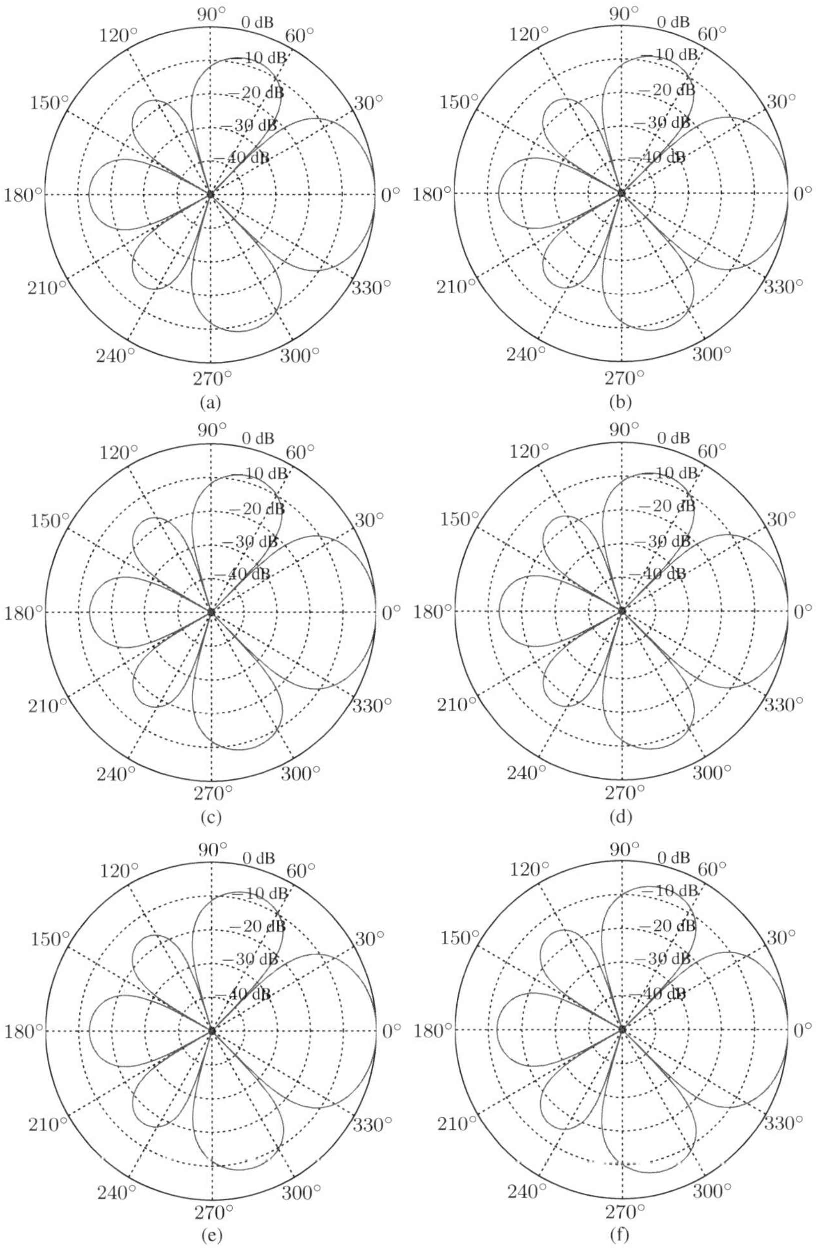

Fig. 5.4 Patterns of Case 3 for several frequencies and two values of δ: (a) $f = 0.5$ kHz, $\delta = 1$ cm, (b) $f = 0.5$ kHz, $\delta = 2$ cm, (c) $f = 1$ kHz, $\delta = 1$ cm, (d) $f = 1$ kHz, $\delta = 2$ cm, (e) $f = 3$ kHz, $\delta = 1$ cm, and (f) $f = 3$ kHz, $\delta = 2$ cm.

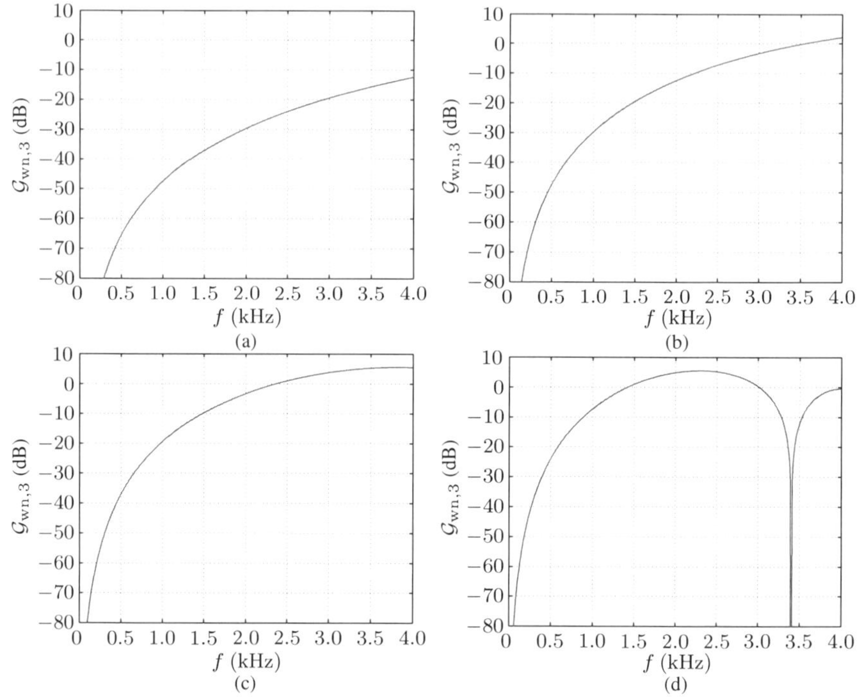

Fig. 5.5 The white noise gain of Case 1, as a function of frequency, for different values of δ: (a) $\delta = 1$ cm, (b) $\delta = 2$ cm, (c) $\delta = 3$ cm, and (d) $\delta = 5$ cm.

We deduce that the white noise is amplified if

$$\omega\tau_0 < \frac{12^{1/6}}{\left[(1-\alpha_{3,1})(1-\alpha_{3,2})(1-\alpha_{3,3})\right]^{1/3}}. \tag{5.11}$$

The directivity factor of the third-order DMA is

$$\mathcal{G}_{dn,3}\left[\mathbf{h}'(\omega)\right] = \frac{\frac{1}{\omega^6}\prod_{n=1}^{3}\left|1 - e^{\jmath\omega\tau_0(1-\alpha_{3,n})}\right|^2}{\mathbf{h}'^H(\omega)\,\mathbf{\Gamma}_{dn}(\omega)\,\mathbf{h}'(\omega)}, \tag{5.12}$$

where

$$\mathbf{\Gamma}_{dn}(\omega) = \begin{bmatrix} 1 & \text{sinc}(\omega\tau_0) & \text{sinc}(2\omega\tau_0) & \text{sinc}(3\omega\tau_0) \\ \text{sinc}(\omega\tau_0) & 1 & \text{sinc}(\omega\tau_0) & \text{sinc}(2\omega\tau_0) \\ \text{sinc}(2\omega\tau_0) & \text{sinc}(\omega\tau_0) & 1 & \text{sinc}(\omega\tau_0) \\ \text{sinc}(3\omega\tau_0) & \text{sinc}(2\omega\tau_0) & \text{sinc}(\omega\tau_0) & 1 \end{bmatrix}. \tag{5.13}$$

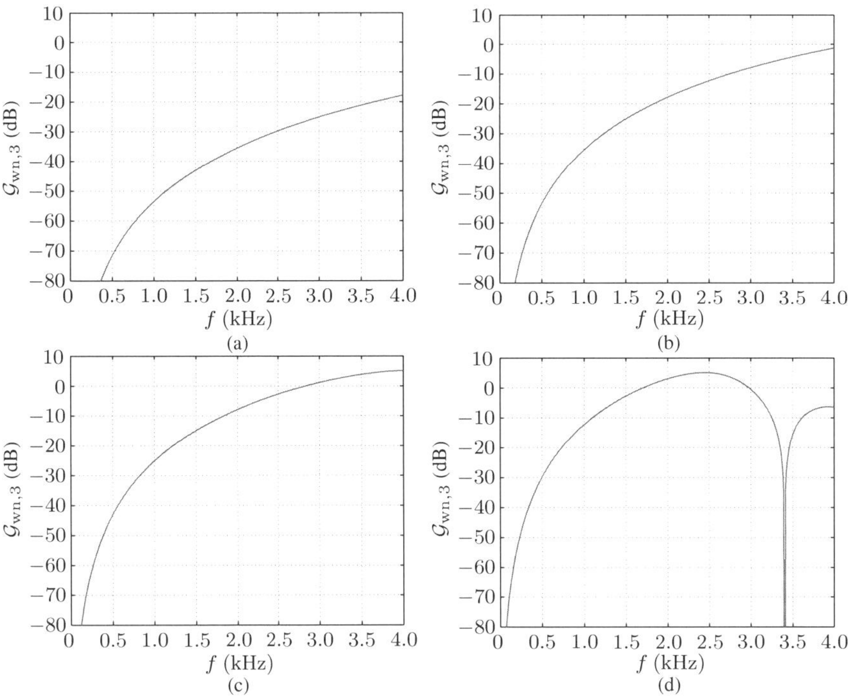

Fig. 5.6 The white noise gain of Case 2, as a function of frequency, for different values of δ: (a) $\delta = 1$ cm, (b) $\delta = 2$ cm, (c) $\delta = 3$ cm, and (d) $\delta = 5$ cm.

In Figs 5.8, 5.9, and 5.10, we plot $\mathcal{G}_{\text{dn},3}\left[\mathbf{h}'(\omega)\right]$ from (5.12), of Case 1, Case 2, and Case 3, respectively, as a function of frequency, for different values of δ.

Finally, it is easy to see that the gain for a point noise source is

$$\mathcal{G}_{\text{ns},3}\left[\mathbf{h}'(\omega)\right] = \frac{\prod_{n=1}^{3}\left\{1 - \cos\left[\omega\tau_0\left(1 - \alpha_{3,n}\right)\right]\right\}}{\prod_{n=1}^{3}\left\{1 - \cos\left[\omega\tau_0\left(\cos\theta_{\text{n}} - \alpha_{3,n}\right)\right]\right\}}. \quad (5.14)$$

Figures 5.11, 5.12, and 5.13 display $\mathcal{G}_{\text{ns},3}\left[\mathbf{h}'(\omega)\right]$, of Case 1, Case 2, and Case 3, respectively, as a function of θ_{n}, for several frequencies and two values of δ.

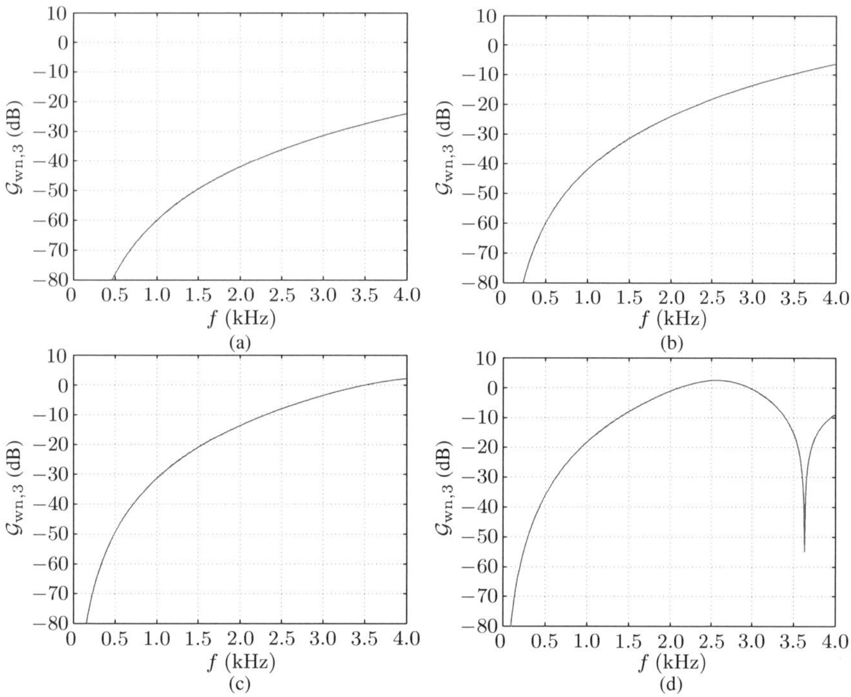

Fig. 5.7 The white noise gain of Case 3, as a function of frequency, for different values of δ: (a) $\delta = 1$ cm, (b) $\delta = 2$ cm, (c) $\delta = 3$ cm, and (d) $\delta = 5$ cm.

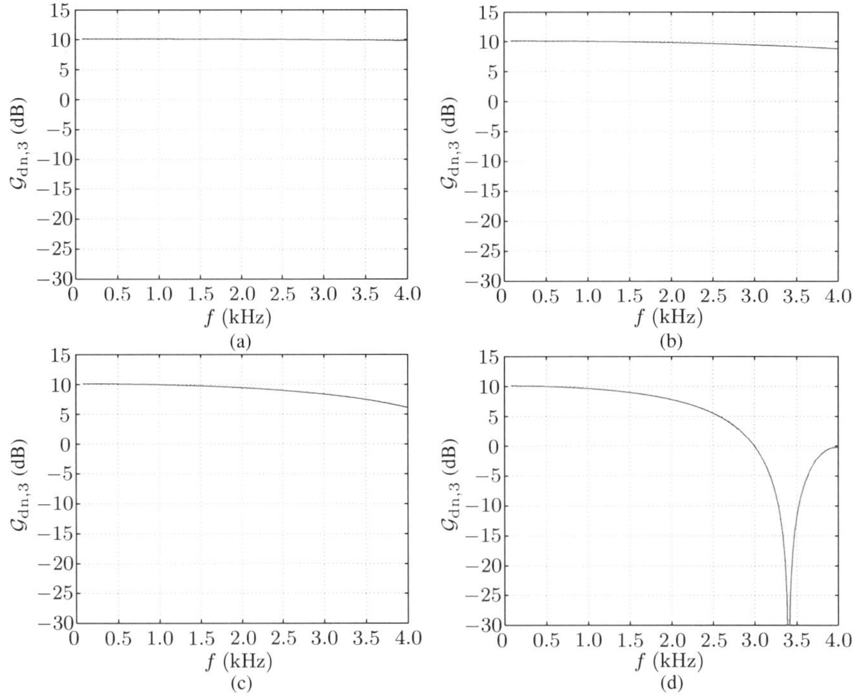

Fig. 5.8 The directivity factor of Case 1, as a function of frequency, for different values of δ: (a) $\delta = 1$ cm, (b) $\delta = 2$ cm, (c) $\delta = 3$ cm, and (d) $\delta = 5$ cm.

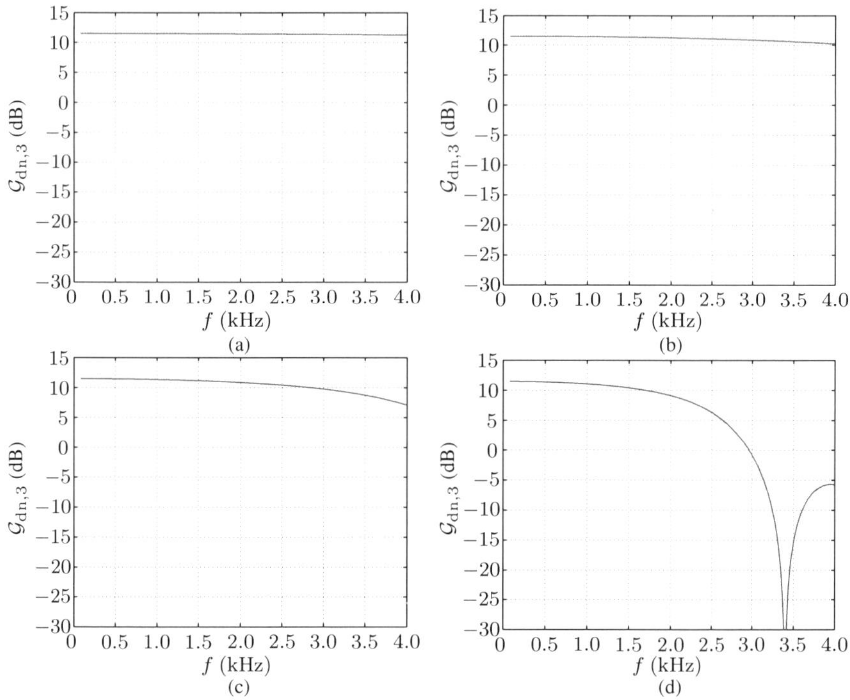

Fig. 5.9 The directivity factor of Case 2, as a function of frequency, for different values of δ: (a) $\delta = 1$ cm, (b) $\delta = 2$ cm, (c) $\delta = 3$ cm, and (d) $\delta = 5$ cm.

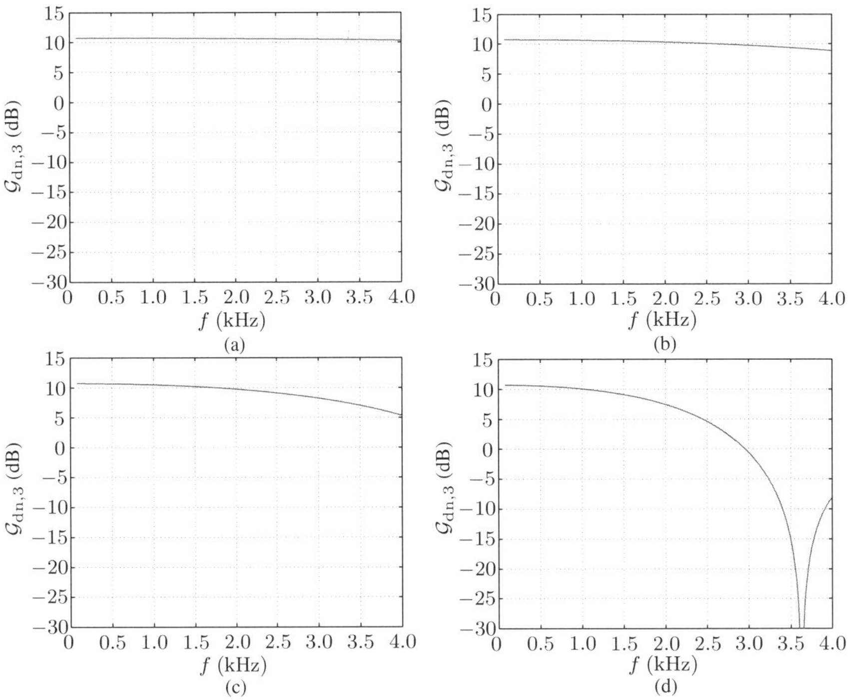

Fig. 5.10 The directivity factor of Case 3, as a function of frequency, for different values of δ: (a) $\delta = 1$ cm, (b) $\delta = 2$ cm, (c) $\delta = 3$ cm, and (d) $\delta = 5$ cm.

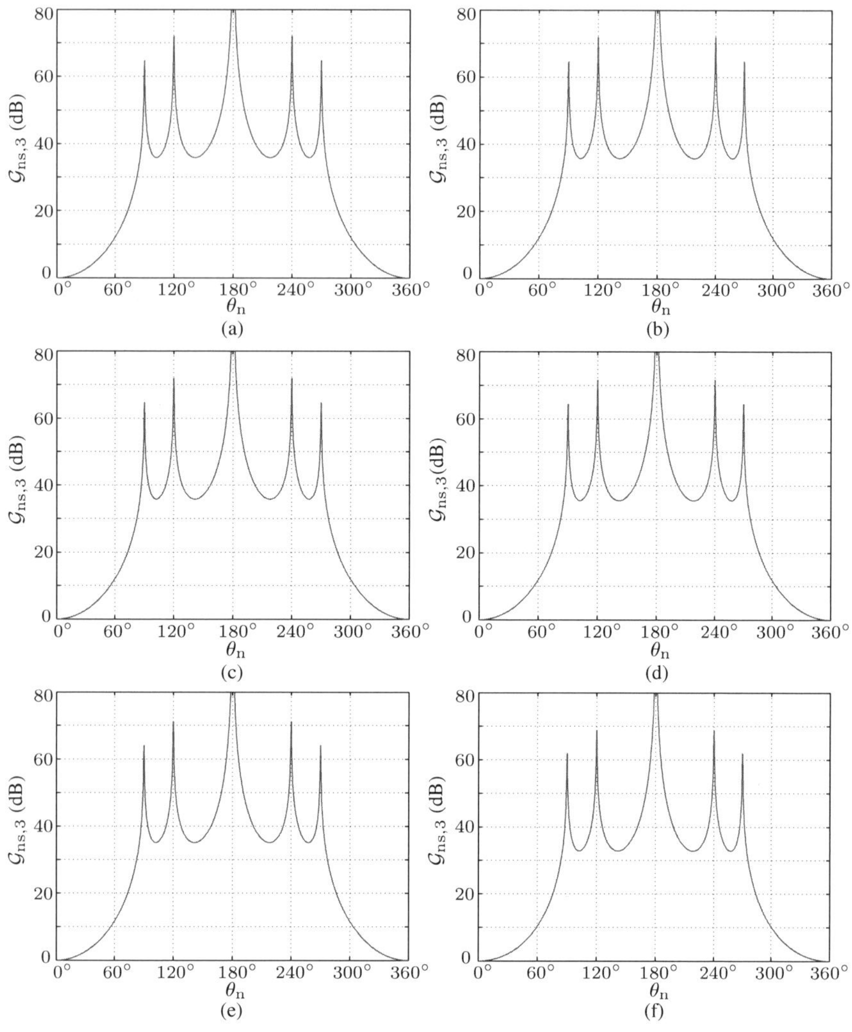

Fig. 5.11 The gain of Case 1 for a point noise source, as a function of θ_{n}, for several frequencies and two values of δ: (a) $f = 0.5$ kHz, $\delta = 1$ cm, (b) $f = 0.5$ kHz, $\delta = 2$ cm, (c) $f = 1$ kHz, $\delta = 1$ cm, (d) $f = 1$ kHz, $\delta = 2$ cm, (e) $f = 3$ kHz, $\delta = 1$ cm, and (f) $f = 3$ kHz, $\delta = 2$ cm.

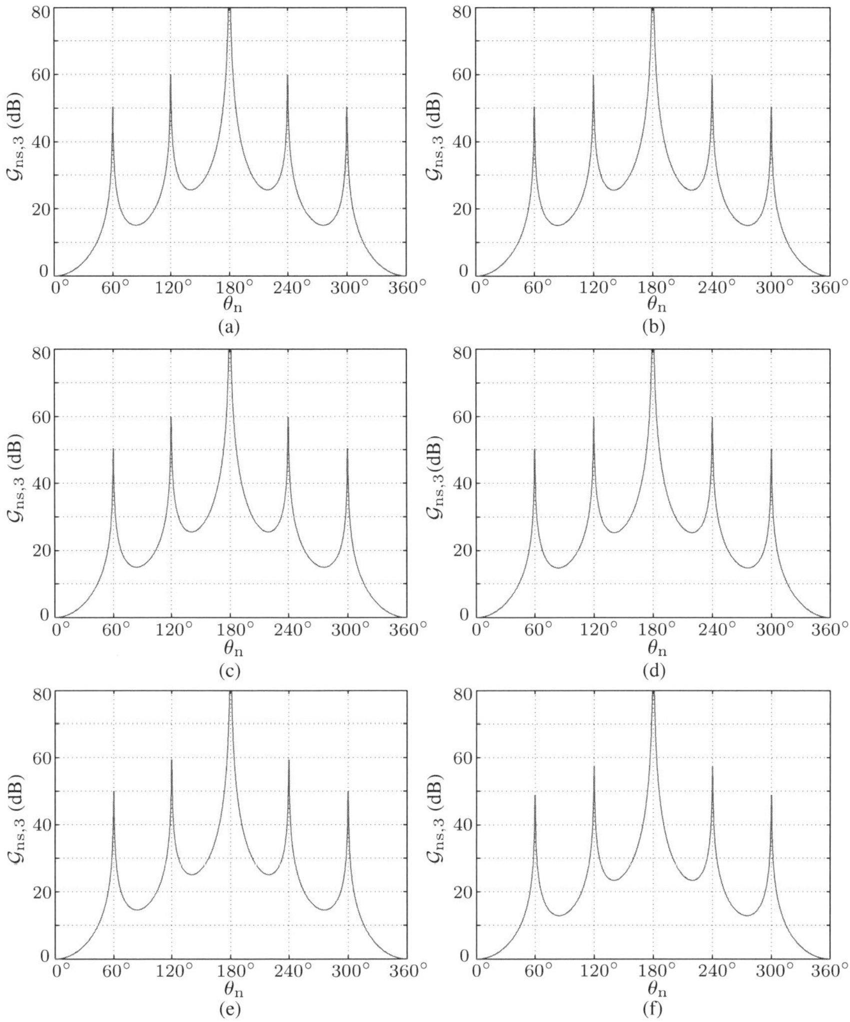

Fig. 5.12 The gain of Case 2 for a point noise source, as a function of θ_n, for several frequencies and two values of δ: (a) $f = 0.5$ kHz, $\delta = 1$ cm, (b) $f = 0.5$ kHz, $\delta = 2$ cm, (c) $f = 1$ kHz, $\delta = 1$ cm, (d) $f = 1$ kHz, $\delta = 2$ cm, (e) $f = 3$ kHz, $\delta = 1$ cm, and (f) $f = 3$ kHz, $\delta = 2$ cm.

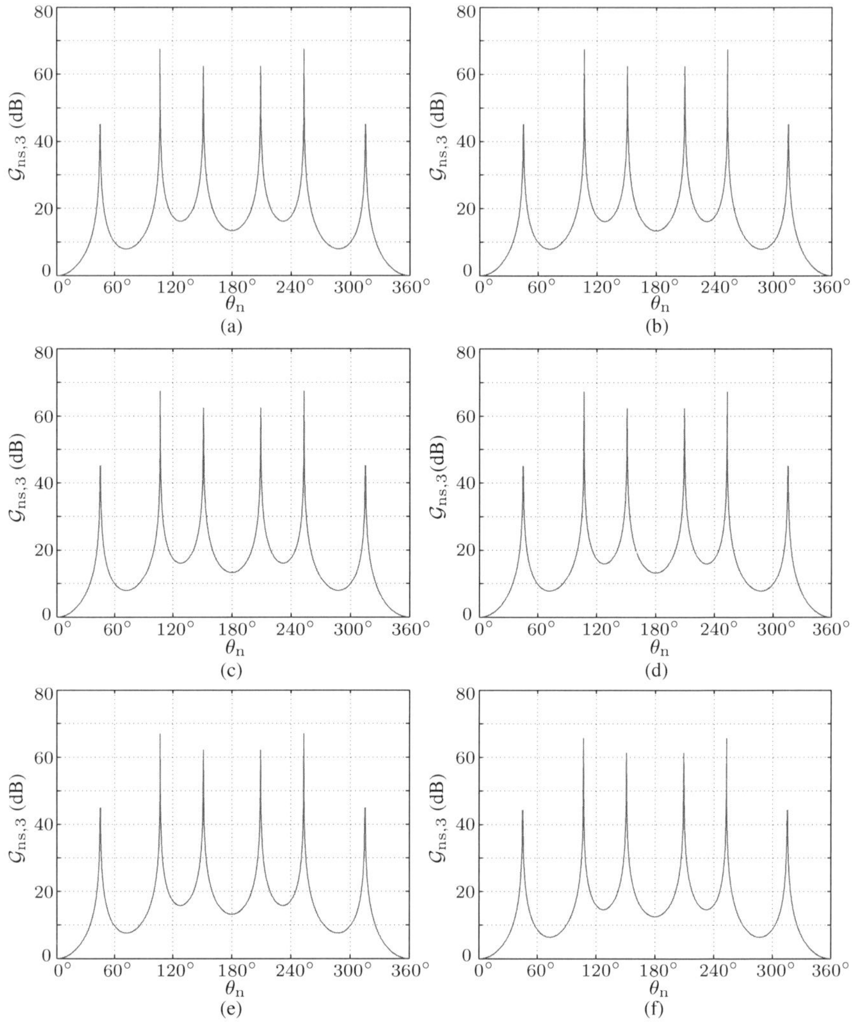

Fig. 5.13 The gain of Case 3 for a point noise source, as a function of θ_{n}, for several frequencies and two values of δ: (a) $f = 0.5$ kHz, $\delta = 1$ cm, (b) $f = 0.5$ kHz, $\delta = 2$ cm, (c) $f = 1$ kHz, $\delta = 1$ cm, (d) $f = 1$ kHz, $\delta = 2$ cm, (e) $f = 3$ kHz, $\delta = 1$ cm, and (f) $f = 3$ kHz, $\delta = 2$ cm.

5.3 Generalization

In this section, we generalize the ideas of the third-order DMA to the Nth order. We seek to generate a directional pattern of order N, which has N distinct nulls, with a uniform linear array of $N+1=M$ microphones. As a result, the linear system of $N+1$ equations to generate this general pattern is

$$\begin{bmatrix} \mathbf{d}^H(\omega,1) \\ \mathbf{d}^H(\omega,\alpha_{N,1}) \\ \mathbf{d}^H(\omega,\alpha_{N,2}) \\ \vdots \\ \mathbf{d}^H(\omega,\alpha_{N,N}) \end{bmatrix} \mathbf{h}(\omega) = \begin{bmatrix} 1 \\ 0 \\ 0 \\ \vdots \\ 0 \end{bmatrix}, \tag{5.15}$$

where $-1 \leq \alpha_{N,1}, \alpha_{N,2}, \ldots, \alpha_{N,N} < 1$ and $\alpha_{N,1} \neq \alpha_{N,2} \neq \cdots \neq \alpha_{N,N}$.

It can be shown that the equivalent filter deduced from (5.15) is

$$\mathbf{h}'^*(\omega) = \frac{1}{\omega^N}\left[1 \; h_2'^*(\omega) \; h_3'^*(\omega) \; \cdots \; h_{N+1}'^*(\omega)\right]^T, \tag{5.16}$$

where

$$\begin{aligned} h_2'^*(\omega) &= -\sum_{i_1=1}^{N} e^{\jmath\omega\tau_0\alpha_{N,i_1}}, \\ h_3'^*(\omega) &= \sum_{i_1\neq i_2} e^{\jmath\omega\tau_0\left(\alpha_{N,i_1}+\alpha_{N,i_2}\right)}, \\ h_4'^*(\omega) &= -\sum_{i_1\neq i_2\neq i_3} e^{\jmath\omega\tau_0\left(\alpha_{N,i_1}+\alpha_{N,i_2}+\alpha_{N,i_3}\right)}, \\ &\vdots \\ h_{n+1}'^*(\omega) &= (-1)^n \sum_{i_1\neq\cdots\neq i_n} e^{\jmath\omega\tau_0\left(\alpha_{N,i_1}+\cdots+\alpha_{N,i_n}\right)}, \\ &\vdots \\ h_{N+1}'^*(\omega) &= (-1)^N e^{\jmath\omega\tau_0\left(\alpha_{N,i_1}+\cdots+\alpha_{N,i_N}\right)}. \end{aligned}$$

This beamformer can be implemented as a cascade of $\frac{N(N+1)}{2}$ first-order DMAs [1]. Thus, the first stage has N first-order DMAs, the second stage has $N-1$ first-order DMAs, etc., and the last stage has one first-order DMA.

The beampattern of the beamformer $\mathbf{h}'(\omega)$ is

$$\mathcal{B}\left[\mathbf{h}'(\omega),\theta\right] = \frac{1}{\omega^N}\prod_{n=1}^{N}\left[1 - e^{\jmath\omega\tau_0\left(\cos\theta-\alpha_{N,n}\right)}\right], \tag{5.17}$$

which can be approximated as the frequency-independent Nth-order directional pattern

$$\mathcal{B}\left[\mathbf{h}'(\omega), \theta\right] \approx \frac{1}{\mathcal{C}^N} \prod_{n=1}^{N} \left(\cos\theta - \alpha_{N,n}\right). \tag{5.18}$$

We can show that the white noise gain is

$$\mathcal{G}_{\mathrm{wn},N}\left[\mathbf{h}'(\omega)\right] = \frac{\prod_{n=1}^{N} \left\{1 - \cos\left[\omega\tau_0 \left(1 - \alpha_{N,n}\right)\right]\right\}}{1 + \frac{1}{2}\sum_{i}\sum_{j>i} \cos\left[\omega\tau_0 \left(\alpha_{N,j} - \alpha_{N,i}\right)\right]}, \tag{5.19}$$

where $\sum_{j>i} \cos\left[\omega\tau_0 \left(\alpha_{N,j} - \alpha_{N,i}\right)\right] = 0$ for $N = 1$. For $N > 1$, we can approximate (5.19) as

$$\mathcal{G}_{\mathrm{wn},N}\left[\mathbf{h}'(\omega)\right] \approx \frac{\left(\omega\tau_0\right)^{2N} \prod_{n=1}^{N} \left(1 - \alpha_{N,n}\right)^2}{3 \times 2^{N-1}}. \tag{5.20}$$

We deduce that the white noise is amplified if

$$\omega\tau_0 < \frac{\left(3 \times 2^{N-1}\right)^{\frac{1}{2N}}}{\left[\prod_{n=1}^{N} \left(1 - \alpha_{N,n}\right)\right]^{1/N}}. \tag{5.21}$$

We also find that the directivity factor and the gain for a point noise source are, respectively,

$$\mathcal{G}_{\mathrm{dn},N}\left[\mathbf{h}'(\omega)\right] = \frac{\frac{1}{\omega^{2N}} \prod_{n=1}^{N} \left|1 - e^{\jmath\omega\tau_0 \left(1 - \alpha_{N,n}\right)}\right|^2}{\mathbf{h}'^H(\omega)\, \mathbf{\Gamma}_{\mathrm{dn}}(\omega)\, \mathbf{h}'(\omega)} \tag{5.22}$$

and

$$\mathcal{G}_{\mathrm{ns},N}\left[\mathbf{h}'(\omega)\right] = \frac{\prod_{n=1}^{N} \left\{1 - \cos\left[\omega\tau_0 \left(1 - \alpha_{N,n}\right)\right]\right\}}{\prod_{n=1}^{N} \left\{1 - \cos\left[\omega\tau_0 \left(\cos\theta_{\mathrm{n}} - \alpha_{N,n}\right)\right]\right\}}. \tag{5.23}$$

References

1. G. W. Elko, "Superdirectional microphone arrays," in *Acoustic Signal Processing for Telecommunication*, S. L. Gay and J. Benesty, Eds. Boston, MA: Kluwer Academic Publishers, 2000, Chapter 10, pp. 181–237.
2. T. D. Abhayapala and A. Gupta, "Higher order differential-integral microphone arrays," *J. Acoust. Soc. Am.*, vol. 127, pp. EL227–EL233, May 2010.

Chapter 6
Minimum-Norm Solution for Robust Differential Arrays

In this chapter, we show how the classical adaptive beamforming technique is related to differential arrays. From this important and useful relationship, we then derive a minimum-norm filter for the design of any order differential array, which can be very robust against white noise amplification. This approach exploits the fact that the number of microphones is greater than the order of the DMA plus one. As a result, the more microphones, the more robust is the DMA for a predetermined order. We recall that all known DMAs are designed in such a way that their order plus one is equal to the number of microphones.

6.1 Adaptive Beamforming versus Differential Arrays

A very important question one may ask: how is adaptive beamforming [1] related to differential arrays [2]?

We know from the previous chapters that any DMA of order N can be designed by solving the linear system of $N+1$ equations:

$$\mathbf{D}(\omega, \boldsymbol{\alpha})\,\mathbf{h}(\omega) = \boldsymbol{\beta}, \tag{6.1}$$

where

$$\mathbf{D}(\omega, \boldsymbol{\alpha}) = \begin{bmatrix} \mathbf{d}^H(\omega, 1) \\ \mathbf{d}^H(\omega, \alpha_{N,1}) \\ \vdots \\ \mathbf{d}^H(\omega, \alpha_{N,N}) \end{bmatrix} \tag{6.2}$$

is the constraint matrix of size $(N+1) \times M$, M is the number of microphones,

J. Benesty and J. Chen, *Study and Design of Differential Microphone Arrays*, Springer Topics in Signal Processing 6, DOI: 10.1007/978-3-642-33753-6_6,

$$\mathbf{d}(\omega, \alpha_{N,n}) = \left[1 \; e^{-\jmath\omega\tau_0\alpha_{N,n}} \; \cdots \; e^{-\jmath(M-1)\omega\tau_0\alpha_{N,n}}\right]^T, \; n = 1, 2, \ldots, N \tag{6.3}$$

is a steering vector of length M,

$$\mathbf{h}(\omega) = \left[H_1(\omega) \; H_2(\omega) \; \cdots \; H_M(\omega)\right]^T \tag{6.4}$$

is a filter of length M, and

$$\boldsymbol{\alpha} = \left[1 \; \alpha_{N,1} \; \cdots \; \alpha_{N,N}\right]^T \tag{6.5}$$

$$\boldsymbol{\beta} = \left[1 \; \beta_{N,1} \; \cdots \; \beta_{N,N}\right]^T \tag{6.6}$$

are vectors of length $N+1$ containing the design coefficients of the directional pattern. In all previous chapters, only the case $M = N+1$ was considered. This is also the case in all known approaches in the literature [2].

In adaptive beamforming, we minimize the residual noise at the beamformer output subject to some constraints. Here, the constraints are summarized in (6.1). Mathematically, this procedure is equivalent to

$$\min_{\mathbf{h}(\omega)} \mathbf{h}^H(\omega)\boldsymbol{\Phi}_{\mathbf{v}}(\omega)\mathbf{h}(\omega) \quad \text{subject to} \quad \mathbf{D}(\omega, \boldsymbol{\alpha})\mathbf{h}(\omega) = \boldsymbol{\beta}, \tag{6.7}$$

for which the solution is

$$\mathbf{h}_{\text{LCMV}}(\omega) = \boldsymbol{\Phi}_{\mathbf{v}}^{-1}(\omega)\mathbf{D}^H(\omega, \boldsymbol{\alpha})\left[\mathbf{D}(\omega, \boldsymbol{\alpha})\boldsymbol{\Phi}_{\mathbf{v}}^{-1}(\omega)\mathbf{D}^H(\omega, \boldsymbol{\alpha})\right]^{-1}\boldsymbol{\beta}, \tag{6.8}$$

where we recognize the well-known linearly constrained minimum variance (LCMV) filter [3], [4], [5]. We observe that for the matrix $\mathbf{D}(\omega, \boldsymbol{\alpha})\boldsymbol{\Phi}_{\mathbf{v}}^{-1}(\omega)\mathbf{D}^H(\omega, \boldsymbol{\alpha})$ in (6.8) to be full rank, we must have $N+1 \leq M$, which is the same condition to design a differential array of order N. It can also be shown that (6.8) can be expressed as

$$\mathbf{h}_{\text{LCMV}}(\omega) = \boldsymbol{\Phi}_{\mathbf{y}}^{-1}(\omega)\mathbf{D}^H(\omega, \boldsymbol{\alpha})\left[\mathbf{D}(\omega, \boldsymbol{\alpha})\boldsymbol{\Phi}_{\mathbf{y}}^{-1}(\omega)\mathbf{D}^H(\omega, \boldsymbol{\alpha})\right]^{-1}\boldsymbol{\beta}, \tag{6.9}$$

where $\boldsymbol{\Phi}_{\mathbf{y}}(\omega) = E\left[\mathbf{y}(\omega)\mathbf{y}^H(\omega)\right]$ is the correlation matrix of the observation signal vector, $\mathbf{y}(\omega)$. However, from an implementation point of view, it is preferable to use (6.8) than (6.9) since $\boldsymbol{\Phi}_{\mathbf{v}}(\omega)$ is usually better conditioned than $\boldsymbol{\Phi}_{\mathbf{y}}(\omega)$.

For $M = N+1$, we easily deduce from (6.8) [or (6.9)] that

$$\mathbf{h}_{\text{LCMV}}(\omega) = \mathbf{D}^{-1}(\omega, \boldsymbol{\alpha})\boldsymbol{\beta}, \tag{6.10}$$

which corresponds exactly to the filter of an Nth-order DMA or the solution of (6.1), which is a linear system of $N+1$ equations and $N+1$ unknowns.

For $M > N+1$ and spatially white noise, (6.8) becomes

$$\mathbf{h}_{\mathrm{LCMV}}(\omega) = \mathbf{D}^H(\omega, \boldsymbol{\alpha}) \left[\mathbf{D}(\omega, \boldsymbol{\alpha}) \mathbf{D}^H(\omega, \boldsymbol{\alpha})\right]^{-1} \boldsymbol{\beta}, \tag{6.11}$$

which corresponds to the minimum-norm solution of (6.1). This shows that the LCMV filter is fundamentally related to the filter of an Nth-order DMA.

Even though we minimize the residual noise at the beamformer output, there is no guaranty with the LCMV filter that the white noise gain should be greater than 1; this is due to the multiple constraints that need to be fulfilled at the same time. However, by virtue of the inclusion principle [6], the solution for $M > N+1$ is better than the solution for $M = N+1$ when it comes to white noise amplification. In fact, the greater is M, the better is the solution. Therefore, contrary to what it may be said in the literature, we can build a high-order DMA without amplifying much the white noise by having more microphones as compared to the order of the DMA.

We can continue our comparisons to the minimum variance distortionless response (MVDR) filter [7], [8]:

$$\mathbf{h}_{\mathrm{MVDR}}(\omega) = \frac{\boldsymbol{\Phi}_{\mathbf{v}}^{-1}(\omega)\, \mathbf{d}(\omega, 1)}{\mathbf{d}^H(\omega, 1)\, \boldsymbol{\Phi}_{\mathbf{v}}^{-1}(\omega)\, \mathbf{d}(\omega, 1)} \tag{6.12}$$

or, equivalently,

$$\mathbf{h}_{\mathrm{MVDR}}(\omega) = \frac{\boldsymbol{\Phi}_{\mathbf{y}}^{-1}(\omega)\, \mathbf{d}(\omega, 1)}{\mathbf{d}^H(\omega, 1)\, \boldsymbol{\Phi}_{\mathbf{y}}^{-1}(\omega)\, \mathbf{d}(\omega, 1)}. \tag{6.13}$$

This filter reduces more noise than the LCMV filter and the gain in SNR is always greater than or equal to 1 (for all types of noise) [9] but $\boldsymbol{\Phi}_{\mathbf{v}}(\omega)$ or $\boldsymbol{\Phi}_{\mathbf{y}}(\omega)$ needs to be estimated accurately; otherwise, the estimation errors may lead to some desired signal cancellation. Moreover, the pattern of the MVDR filter does not correspond, in general, to any known directional pattern contrary to the LCMV filter. The only exception is when $\boldsymbol{\Phi}_{\mathbf{v}}(\omega)$ in (6.12) is replaced by $\boldsymbol{\Gamma}_{\mathrm{dn}}(\omega)$. In this case, indeed, the MVDR filter corresponds to the pattern of the hypercardioid (of order $M-1$) but the white noise can be highly amplified since when the number of microphones is increased so is the order of the hypercardioid. This optimization procedure maximizes the directivity factor at the expense of white noise amplification. While in the LCMV filter, we can increase the number of microphones without increasing much the order of the DMA; as a result, white noise amplification can be controlled. We will get back to this point in the next section.

This study also suggests that adaptive beamforming with a linear array should be greatly better in terms of noise reduction when the desired source signal propagates from the endfire than any other directions. This way, we also take advantage of the nature of DMAs.

6.2 Minimum-Norm Filter

As it is clearly explained in the previous section, the minimum-norm solution given in (6.11) is of great importance here since it has the potential to solve the problem of white noise amplification encountered in all known differential arrays. Let us slightly change the notation of this filter and denote it by

$$\mathbf{h}(\omega, \boldsymbol{\alpha}, \boldsymbol{\beta}) = \mathbf{D}^H(\omega, \boldsymbol{\alpha}) \left[\mathbf{D}(\omega, \boldsymbol{\alpha}) \mathbf{D}^H(\omega, \boldsymbol{\alpha})\right]^{-1} \boldsymbol{\beta}, \tag{6.14}$$

where, again, the vectors $\boldsymbol{\alpha}$ and $\boldsymbol{\beta}$ of length $N+1$ determine the pattern and the order of the DMA. Basically, the length of these vectors determine (roughly) the order of the DMA while their values determine the pattern. Meanwhile, the length, M, of the minimum-norm filter, $\mathbf{h}(\omega, \boldsymbol{\alpha}, \boldsymbol{\beta})$, can be much larger than $N+1$, which will help make the DMA robust against white noise amplification. In our context, a more rigorous way to derive the minimum-norm filter is by maximizing the white noise gain subject to the Nth-order DMA fundamental constraints. This is equivalent to minimizing $\mathbf{h}^H(\omega, \boldsymbol{\alpha}, \boldsymbol{\beta}) \mathbf{h}(\omega, \boldsymbol{\alpha}, \boldsymbol{\beta})$ subject to (6.1), from which (6.14) results. In other words, the LCMV filter in the presence of white noise is the solution we are looking for. Therefore, for a large number of microphones, the white noise gain should approach M. If the value of M is much larger than $N+1$, the order of the DMA may not be equal to N anymore but the Nth-order DMA fundamental constraints will always be fulfilled. As a result, the resulting shape of the directional pattern may slightly be different than the one obtained with $M = N+1$. This approach is the best we can do to solve the conflicting requirement of a high-order DMA that does not amplify the white noise.

We suggest to use (6.14) for the design of any order DMA. Of course, its implementation is going to be very different from the classical one where only delays and a low-pass filter are needed. Here, instead, we should rely on a subband or frequency-domain implementation, which can still be very efficient from a complexity point of view.

It is easy to see that the beampattern of the minimum-norm filter is

$$\begin{aligned}\mathcal{B}\left[\mathbf{h}(\omega, \boldsymbol{\alpha}, \boldsymbol{\beta}), \theta\right] &= \mathbf{d}^H(\omega, \cos\theta) \mathbf{h}(\omega, \boldsymbol{\alpha}, \boldsymbol{\beta}) \\ &= \mathbf{d}^H(\omega, \cos\theta) \mathbf{D}^H(\omega, \boldsymbol{\alpha}) \left[\mathbf{D}(\omega, \boldsymbol{\alpha}) \mathbf{D}^H(\omega, \boldsymbol{\alpha})\right]^{-1} \boldsymbol{\beta}.\end{aligned} \tag{6.15}$$

We also deduce that the white noise gain, the directivity factor, and the gain for a point noise source are, respectively,

$$\mathcal{G}_{\text{wn}}\left[\mathbf{h}(\omega, \boldsymbol{\alpha}, \boldsymbol{\beta})\right] = \frac{1}{\boldsymbol{\beta}^T \left[\mathbf{D}(\omega, \boldsymbol{\alpha}) \mathbf{D}^H(\omega, \boldsymbol{\alpha})\right]^{-1} \boldsymbol{\beta}}, \tag{6.16}$$

$$\mathcal{G}_{\mathrm{dn}}\left[\mathbf{h}\left(\omega,\boldsymbol{\alpha},\boldsymbol{\beta}\right)\right]=\frac{1}{\mathbf{h}^{H}\left(\omega,\boldsymbol{\alpha},\boldsymbol{\beta}\right)\boldsymbol{\Gamma}_{\mathrm{dn}}\left(\omega\right)\mathbf{h}\left(\omega,\boldsymbol{\alpha},\boldsymbol{\beta}\right)},\tag{6.17}$$

and

$$\mathcal{G}_{\mathrm{ns}}\left[\mathbf{h}\left(\omega,\boldsymbol{\alpha},\boldsymbol{\beta}\right)\right]=\frac{1}{\left|\mathcal{B}\left[\mathbf{h}\left(\omega,\boldsymbol{\alpha},\boldsymbol{\beta}\right),\theta_{\mathrm{n}}\right]\right|^{2}},\tag{6.18}$$

where θ_{n} is the angle of the point noise source (see Chapter 2).

Ideally, we would like to have

$$\mathcal{G}_{\mathrm{dn}}\left[\mathbf{h}\left(\omega,\boldsymbol{\alpha},\boldsymbol{\beta}\right)\right]=\mathcal{G}_{N},\tag{6.19}$$

where $\mathcal{G}_{N}$ is the directivity factor of the frequency-independent Nth-order DMA (see Chapter 2) and

$$\mathcal{G}_{\mathrm{wn}}\left[\mathbf{h}\left(\omega,\boldsymbol{\alpha},\boldsymbol{\beta}\right)\right]\geq 1.\tag{6.20}$$

6.3 Design Examples

In this section, we show how to design several directional patterns of different orders with the minimum-norm filter and a uniform linear array of up to eight microphones. We only study four examples. Evidently, many more patterns can be designed with the proposed approach. In the previous chapters, we considered a bandwidth of 4 kHz; here, we use a bandwidth of 8 kHz in order to have a better insight into this new method. In all examples, the inter-element spacing is $\delta = 1$ cm.

6.3.1 First-Order Cardioid

From Chapter 3, we know that the parameters to design the first-order cardioid are summarized into the two vectors of length 2:

$$\boldsymbol{\alpha}=\left[1\ -1\right]^{T},\tag{6.21}$$

$$\boldsymbol{\beta}=\left[1\ 0\right]^{T}.\tag{6.22}$$

Figures 6.1, 6.2, and 6.3 display the patterns of the first-order cardioid with 2, 5, and 8 microphones, respectively, for several frequencies. The patterns for the 2 and 5 microphones cases are similar except for 5 kHz, where the pattern with 5 microphones is more directional than the one with 2 microphones. With 8 microphones, we clearly see that for 4 and 5 kHz, we get the directional patterns of the second-order cardioid, which is expected and interesting. In this scenario, it is more appropriate to talk about a mixed first/second-order cardioid (first order at low frequencies and second order at high frequencies).

Figure 6.4 gives plots of the white noise gain of the first-order cardioid, as a function of frequency, for different values of M. As M increases, so is the white noise gain. With 7 and 8 microphones and around 2 kHz and above, we see that we reach the maximum white noise gain. Now, if we compare the 2 and 5 microphone cases, we observe that at 1 kHz, the gain is above 0 dB with 5 microphones and around -11 dB with 2 microphones, which represents an improvement of more than 11 dB.

Figure 6.5 shows plots of the directivity factor of the first-order cardioid, as a function of frequency, for different values of M. For $M = 2$, the directivity factor is almost constant up to 4 kHz. As M increases, the frequency range for which the directivity factor is constant decreases but at high frequencies, we can get much better gains than at low frequencies, which is what we already observed in the patterns.

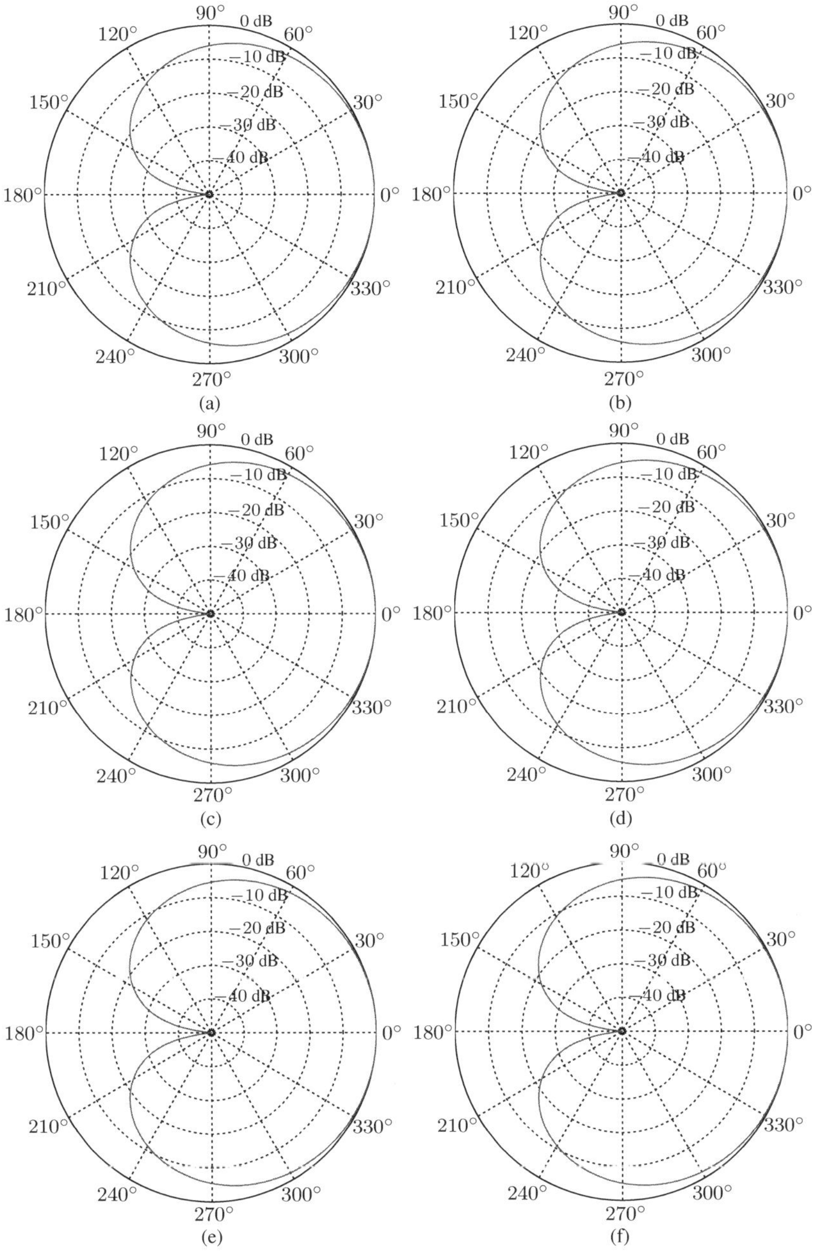

Fig. 6.1 Patterns of the first-order cardioid with 2 microphones for several frequencies: (a) $f = 0.5$ kHz, (b) $f = 1.0$ kHz, (c) $f = 2.0$ kHz, (d) $f = 3.0$ kHz, (e) $f = 4.0$ kHz, and (f) $f = 5.0$ kHz.

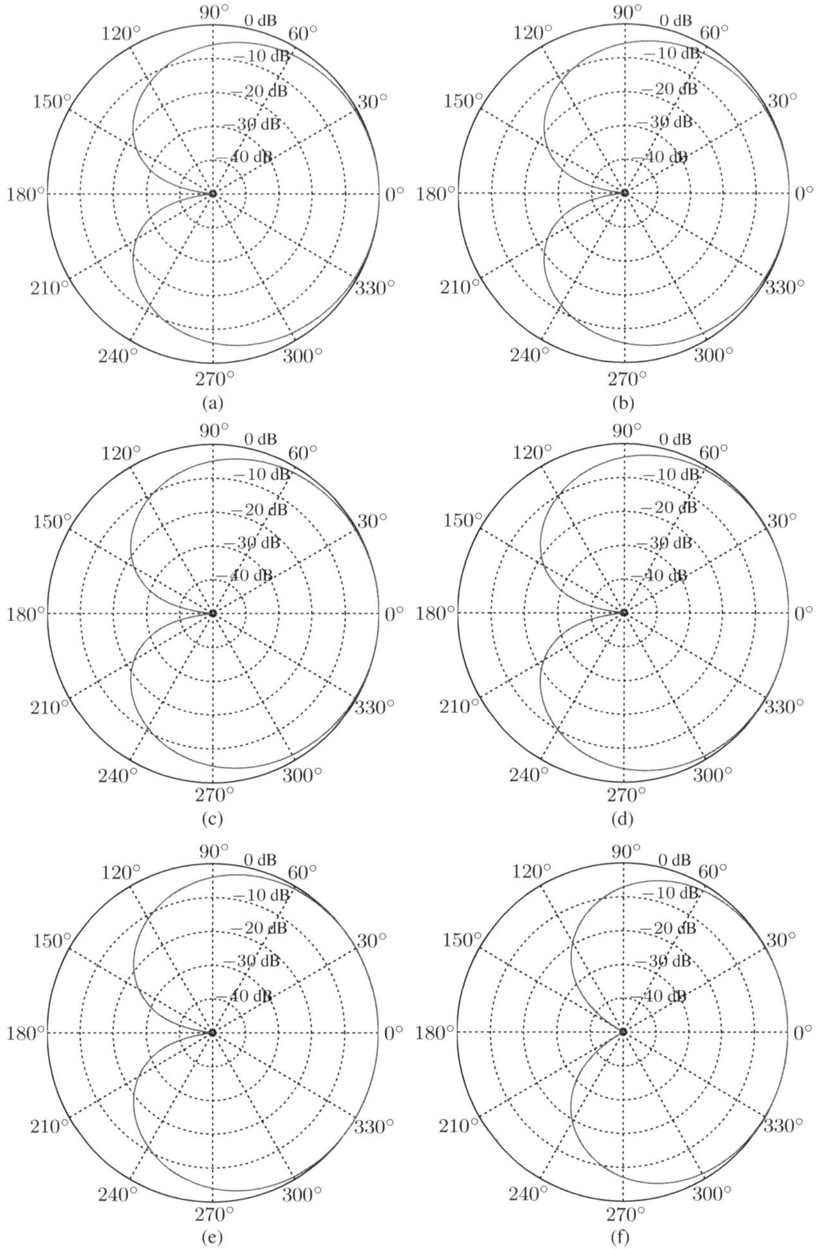

Fig. 6.2 Patterns of the first-order cardioid with 5 microphones for several frequencies: (a) $f = 0.5$ kHz, (b) $f = 1.0$ kHz, (c) $f = 2.0$ kHz, (d) $f = 3.0$ kHz, (e) $f = 4.0$ kHz, and (f) $f = 5.0$ kHz.

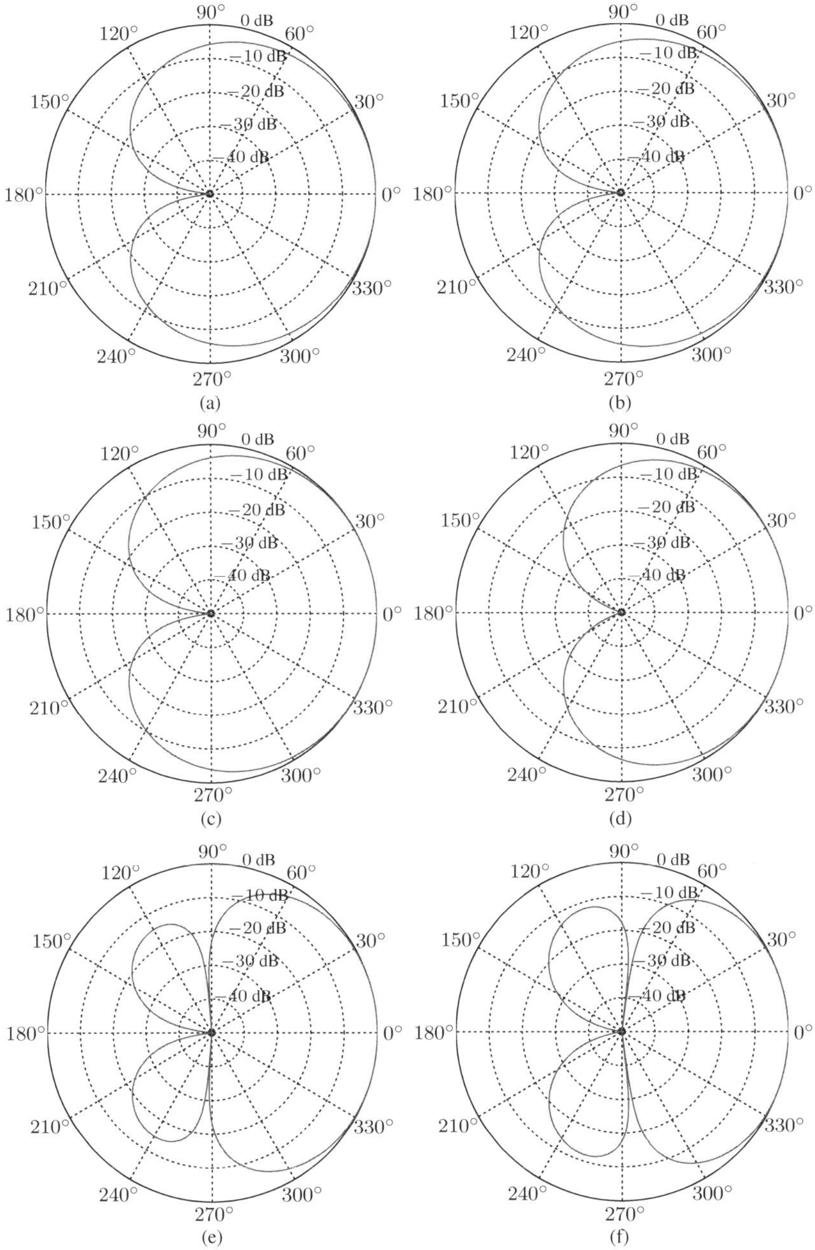

Fig. 6.3 Patterns of the first-order cardioid with 8 microphones for several frequencies: (a) $f = 0.5$ kHz, (b) $f = 1.0$ kHz, (c) $f = 2.0$ kHz, (d) $f = 3.0$ kHz, (e) $f = 4.0$ kHz, and (f) $f = 5.0$ kHz.

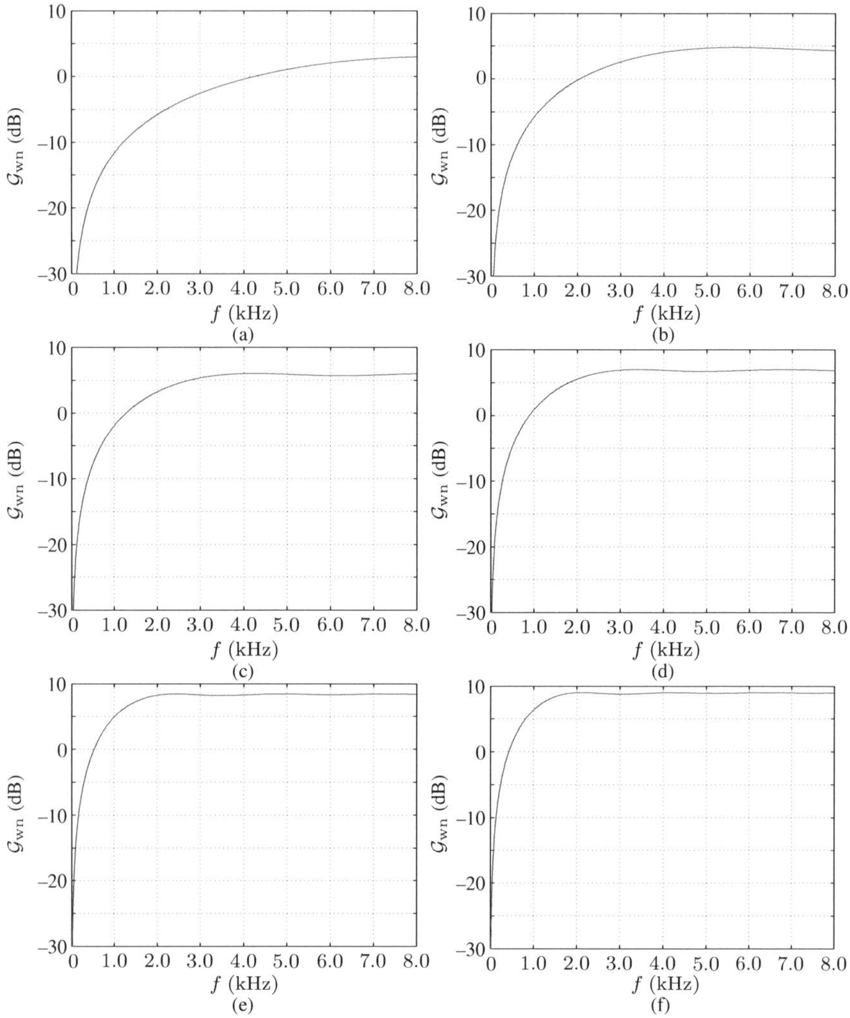

Fig. 6.4 The white noise gain of the first-order cardioid, as a function of frequency, for different values of M: (a) $M = 2$, (b) $M = 3$, (c) $M = 4$, (d) $M = 5$, (e) $M = 7$, and (f) $M = 8$.

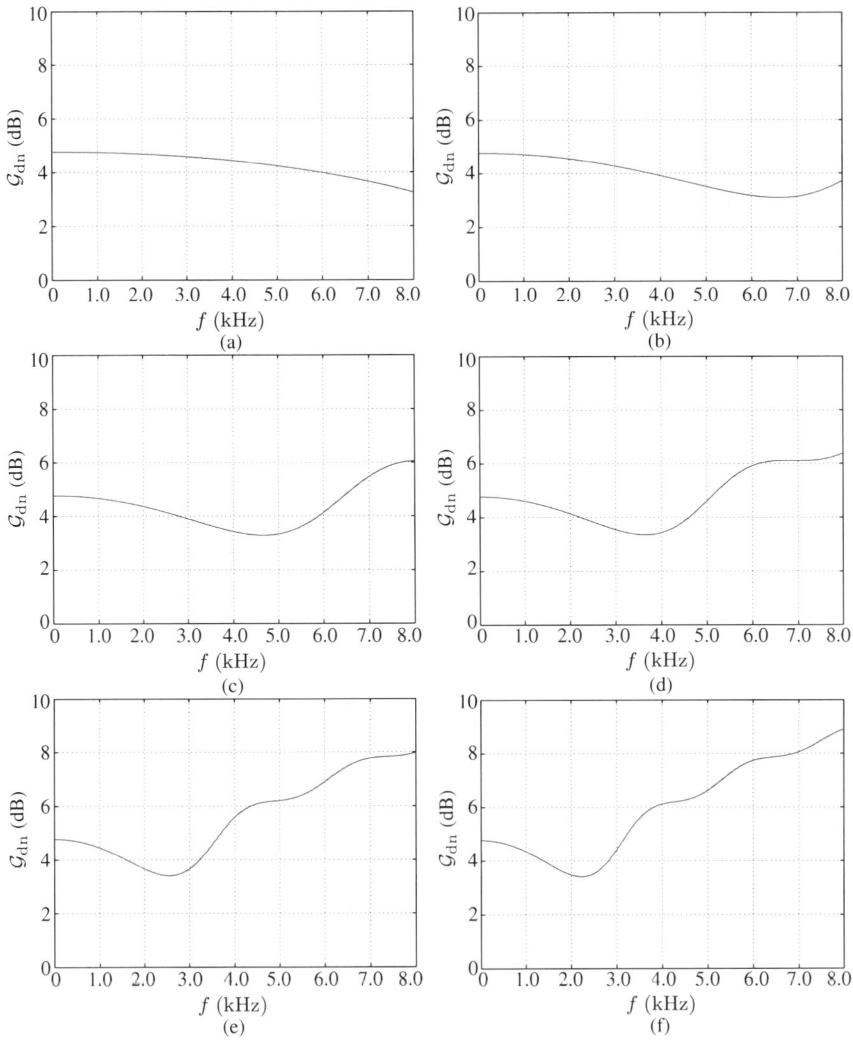

Fig. 6.5 The directivity factor of the first-order cardioid, as a function of frequency, for different values of M: (a) $M = 2$, (b) $M = 3$, (c) $M = 4$, (d) $M = 5$, (e) $M = 7$, and (f) $M = 8$.

6.3.2 Second-Order Cardioid

From Chapter 4 on second-order DMAs, we find that the parameters of the three fundamental constraints for the second-order cardioid are

$$\boldsymbol{\alpha} = \begin{bmatrix} 1 & -1 & 0 \end{bmatrix}^T, \tag{6.23}$$

$$\boldsymbol{\beta} = \begin{bmatrix} 1 & 0 & 0 \end{bmatrix}^T. \tag{6.24}$$

In Figs 6.6, 6.7, and 6.8, we plot the patterns of the second-order cardioid with 3, 5, and 8 microphones, respectively, for several frequencies. We observe some minor differences when the number of microphones increases but all the patterns correspond roughly to the second-order cardioid.

In Fig. 6.9, we give plots of the white noise gain of the second-order cardioid, as a function of frequency, for different values of M. As expected, the white noise gain increases as M increases but the maximum is reached only at high frequencies and for $M \geq 7$. Comparing the 3 and 5 microphone cases at 1 kHz, we observe that the gain is -19 dB with 5 microphones and -30 dB with 3 microphones, which represents an improvement of 11 dB. For $M = 8$, the white noise gain is only -9 dB at 1 kHz.

Figure 6.10 shows plots of the directivity factor of the second-order cardioid, as a function of frequency, for different values of M. Comparing the 3 and 5 microphone cases, we see that the directivity factor is roughly the same in both cases for frequencies up to 4 kHz; for higher frequencies, the gain is a bit worst for $M = 5$.

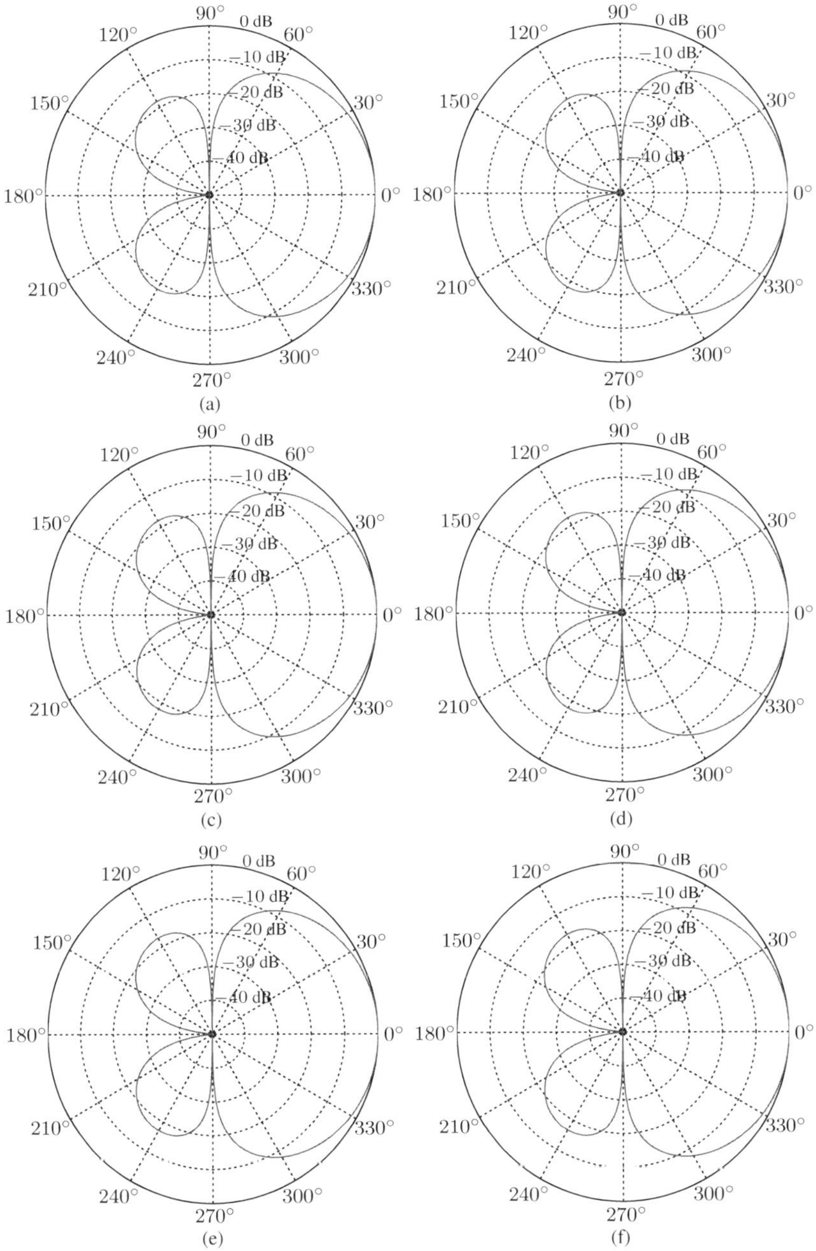

Fig. 6.6 Patterns of the second-order cardioid with 3 microphones for several frequencies: (a) $f = 0.5$ kHz, (b) $f = 1.0$ kHz, (c) $f = 2.0$ kHz, (d) $f = 3.0$ kHz, (e) $f = 4.0$ kHz, and (f) $f = 5.0$ kHz.

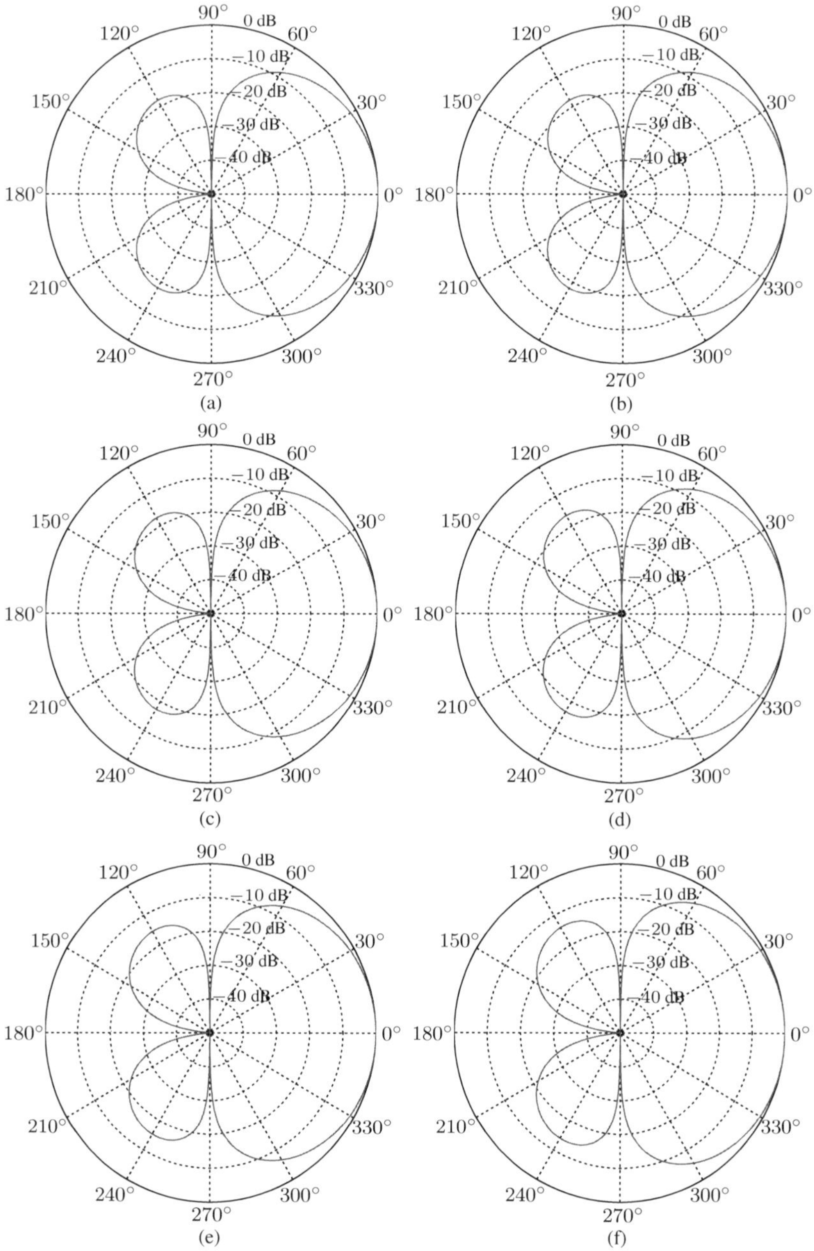

Fig. 6.7 Patterns of the second-order cardioid with 5 microphones for several frequencies: (a) $f = 0.5$ kHz, (b) $f = 1.0$ kHz, (c) $f = 2.0$ kHz, (d) $f = 3.0$ kHz, (e) $f = 4.0$ kHz, and (f) $f = 5.0$ kHz.

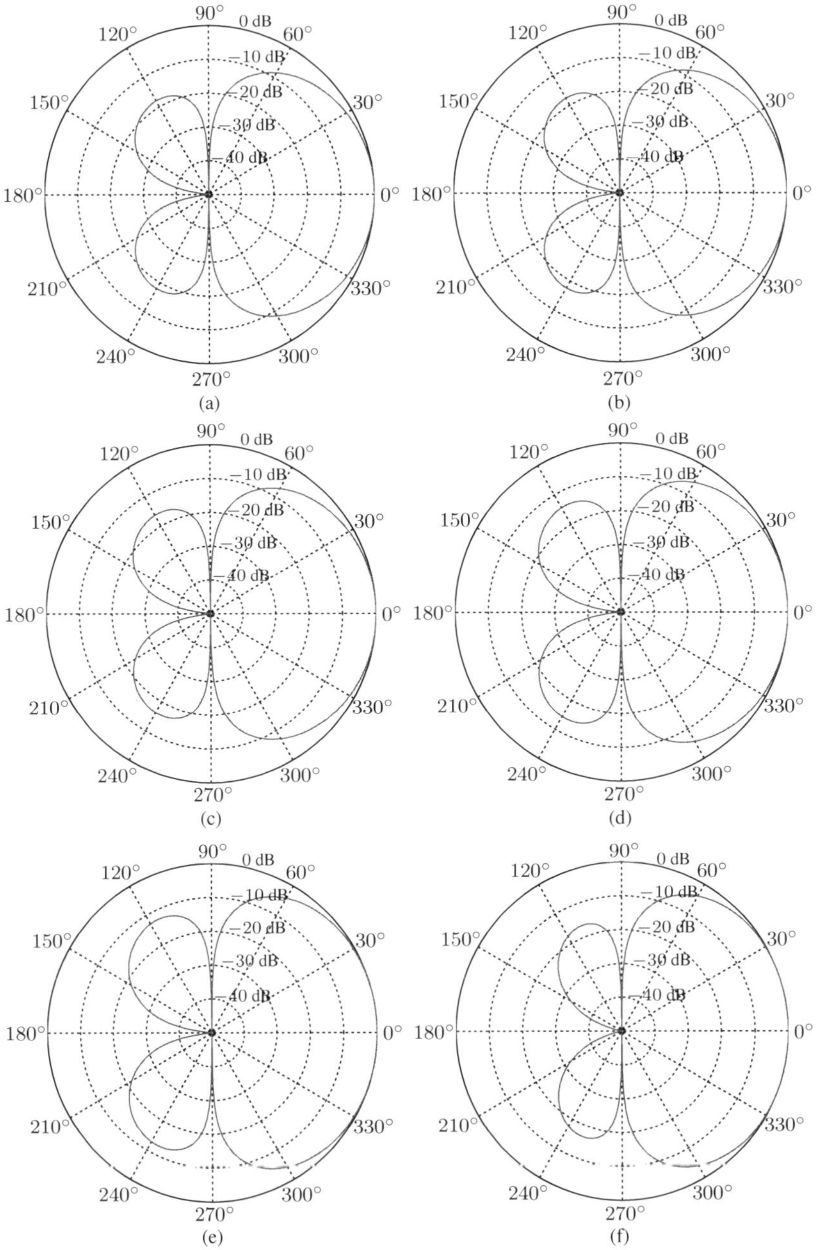

Fig. 6.8 Patterns of the second-order cardioid with 8 microphones for several frequencies: (a) $f = 0.5$ kHz, (b) $f = 1.0$ kHz, (c) $f = 2.0$ kHz, (d) $f = 3.0$ kHz, (e) $f = 4.0$ kHz, and (f) $f = 5.0$ kHz.

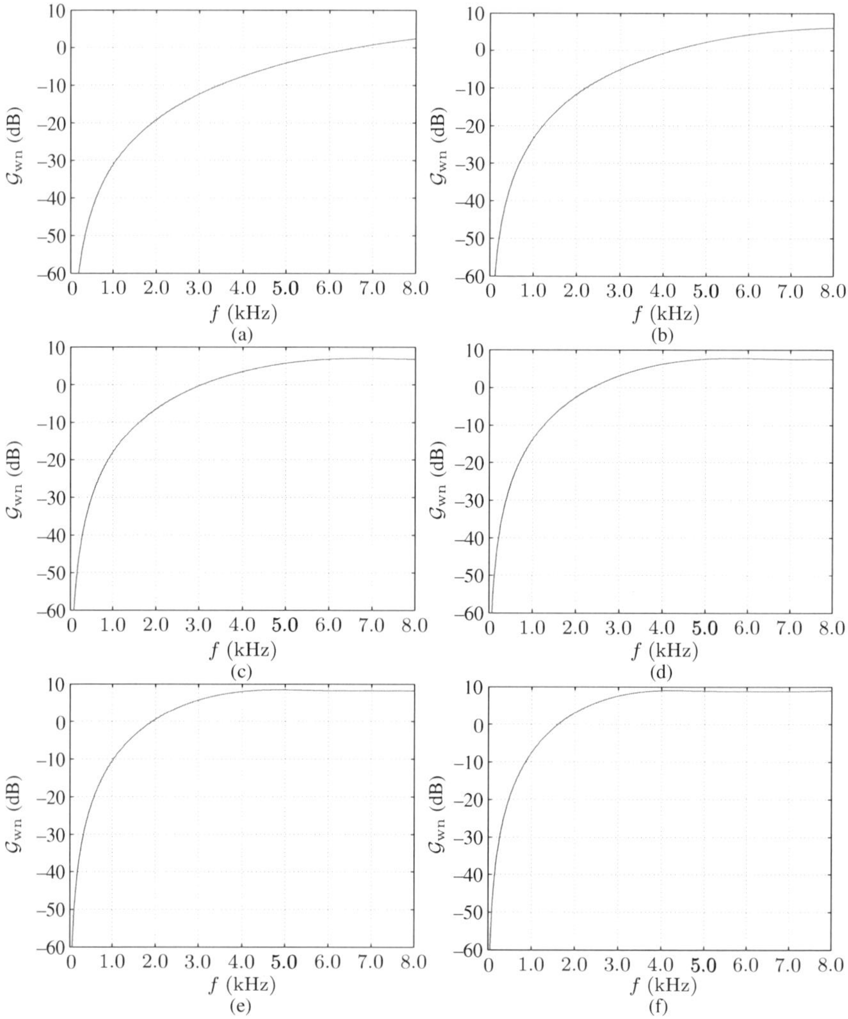

Fig. 6.9 The white noise gain of the second-order cardioid, as a function of frequency, for different values of M: (a) $M = 3$, (b) $M = 4$, (c) $M = 5$, (d) $M = 6$, (e) $M = 7$, and (f) $M = 8$.

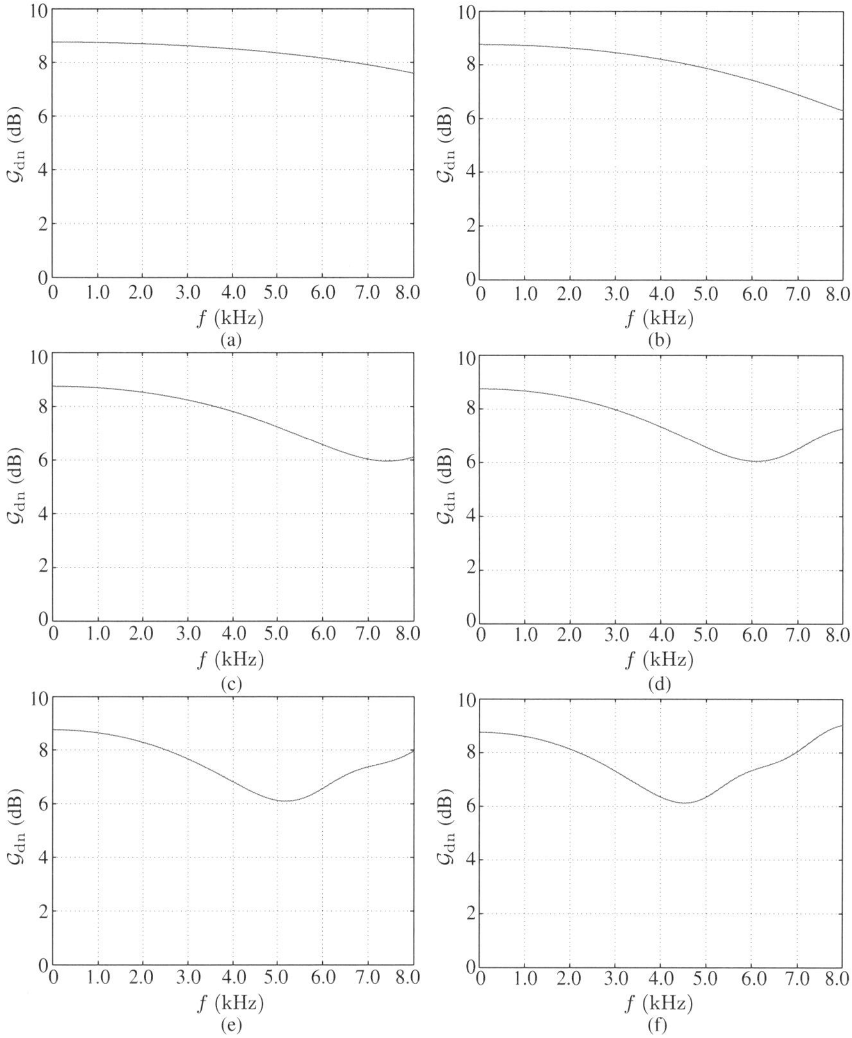

Fig. 6.10 The directivity factor of the second-order cardioid, as a function of frequency, for different values of M: (a) $M = 3$, (b) $M = 4$, (c) $M = 5$, (d) $M = 6$, (e) $M = 7$, and (f) $M = 8$.

6.3.3 Third-Order Cardioid

It can be checked that the parameters of the four fundamental constraints to design a third-order cardioid are summarized into the two vectors:

$$\boldsymbol{\alpha} = \left[1 \ -1 \ 0 \ -\tfrac{\sqrt{2}}{2}\right]^T, \tag{6.25}$$

$$\boldsymbol{\beta} = \left[1 \ 0 \ 0 \ -\tfrac{\sqrt{2}}{8} + \tfrac{1}{4}\right]^T. \tag{6.26}$$

Figures 6.11, 6.12, and 6.13 display the patterns of the third-order cardioid with 4, 5, and 8 microphones, respectively, for several frequencies. All patterns look similar.

In Fig. 6.14, we give plots of the white noise gain of the third-order cardioid, as a function of frequency, for different values of M. Comparing the 4 and 8 microphone cases at 1 kHz, we observe that the gain is -50 dB with 4 microphones and around -24 dB with 8 microphones, which represents an improvement of about 26 dB.

In Fig. 6.15, we show plots of the directivity factor of the third-order cardioid, as a function of frequency, for different values of M. Comparing the 4 and 8 microphone cases, we see that the gain in the second design is only slightly worse in the range 3–6 kHz.

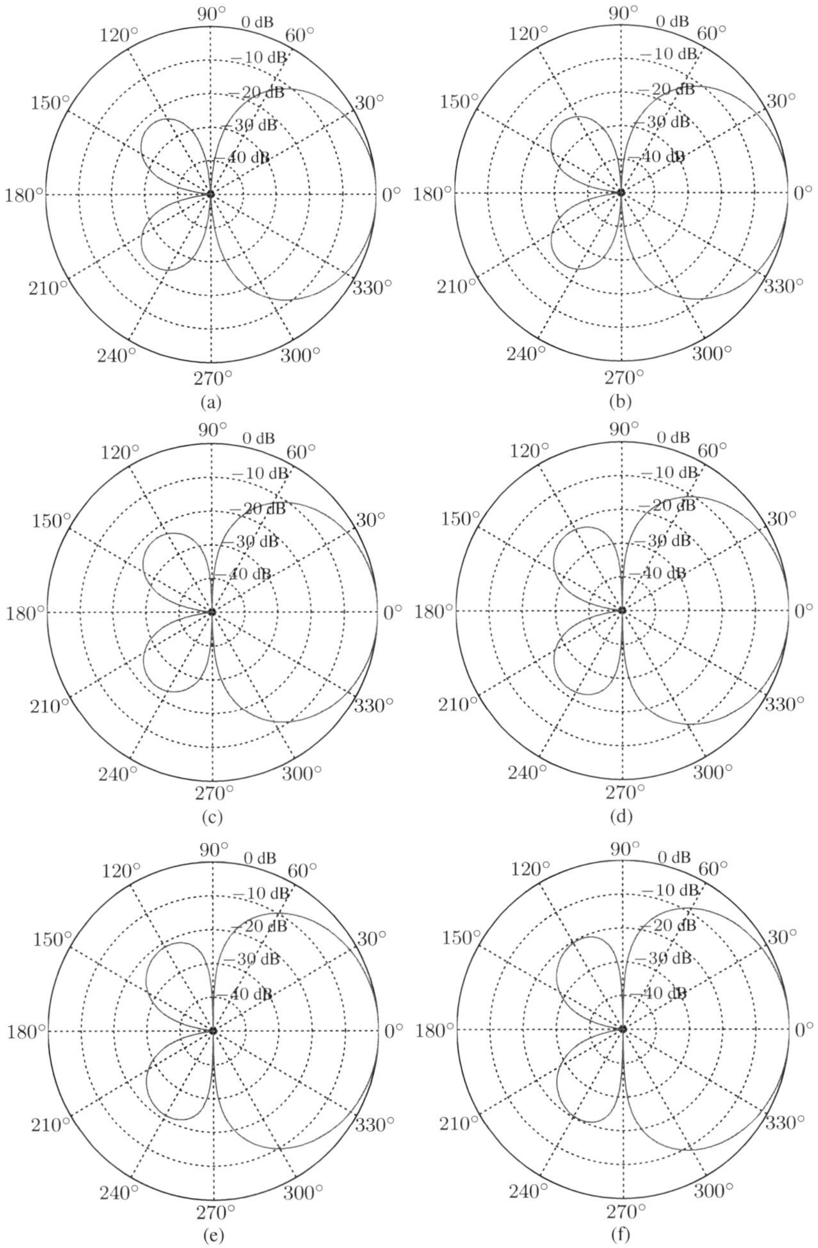

Fig. 6.11 Patterns of the third-order cardioid with 4 microphones for several frequencies: (a) $f = 0.5$ kHz, (b) $f = 1.0$ kHz, (c) $f = 2.0$ kHz, (d) $f = 3.0$ kHz, (e) $f = 4.0$ kHz, and (f) $f = 5.0$ kHz.

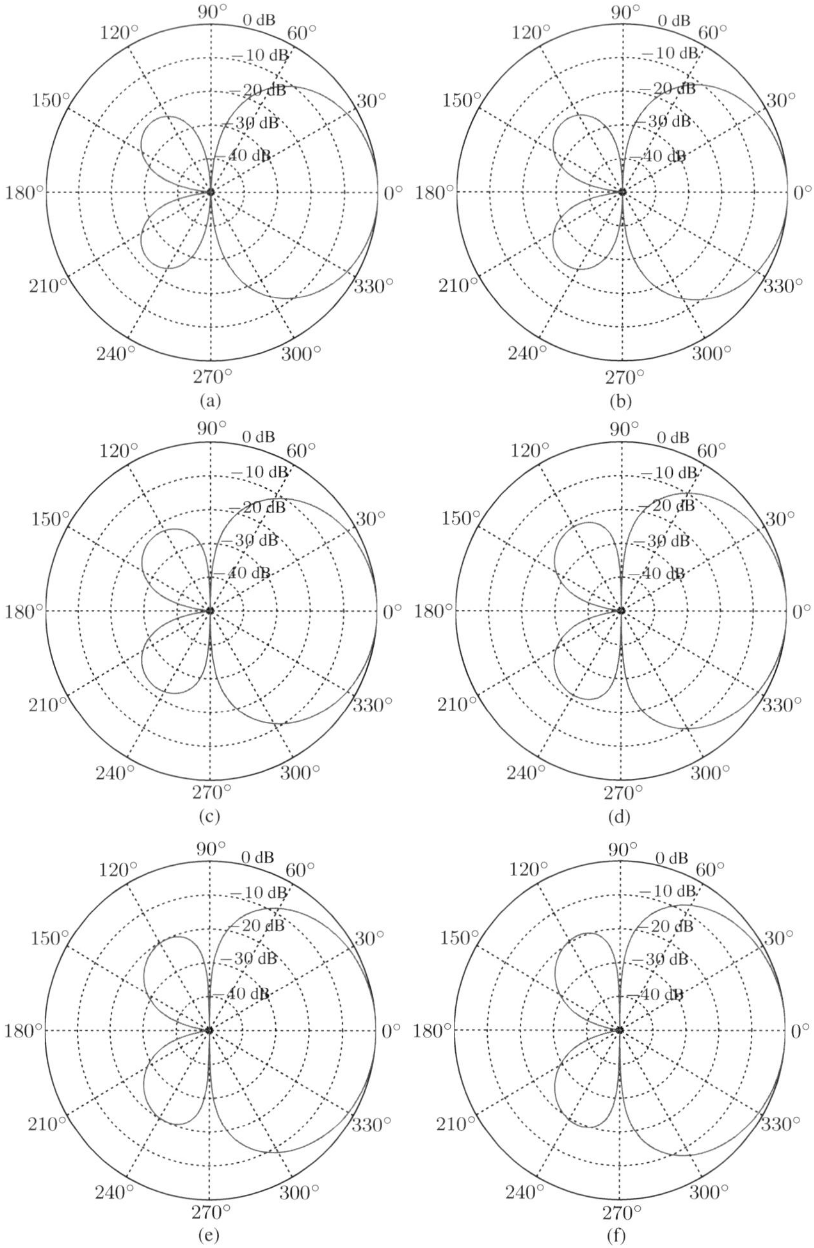

Fig. 6.12 Patterns of the third-order cardioid with 5 microphones for several frequencies: (a) $f = 0.5$ kHz, (b) $f = 1.0$ kHz, (c) $f = 2.0$ kHz, (d) $f = 3.0$ kHz, (e) $f = 4.0$ kHz, and (f) $f = 5.0$ kHz.

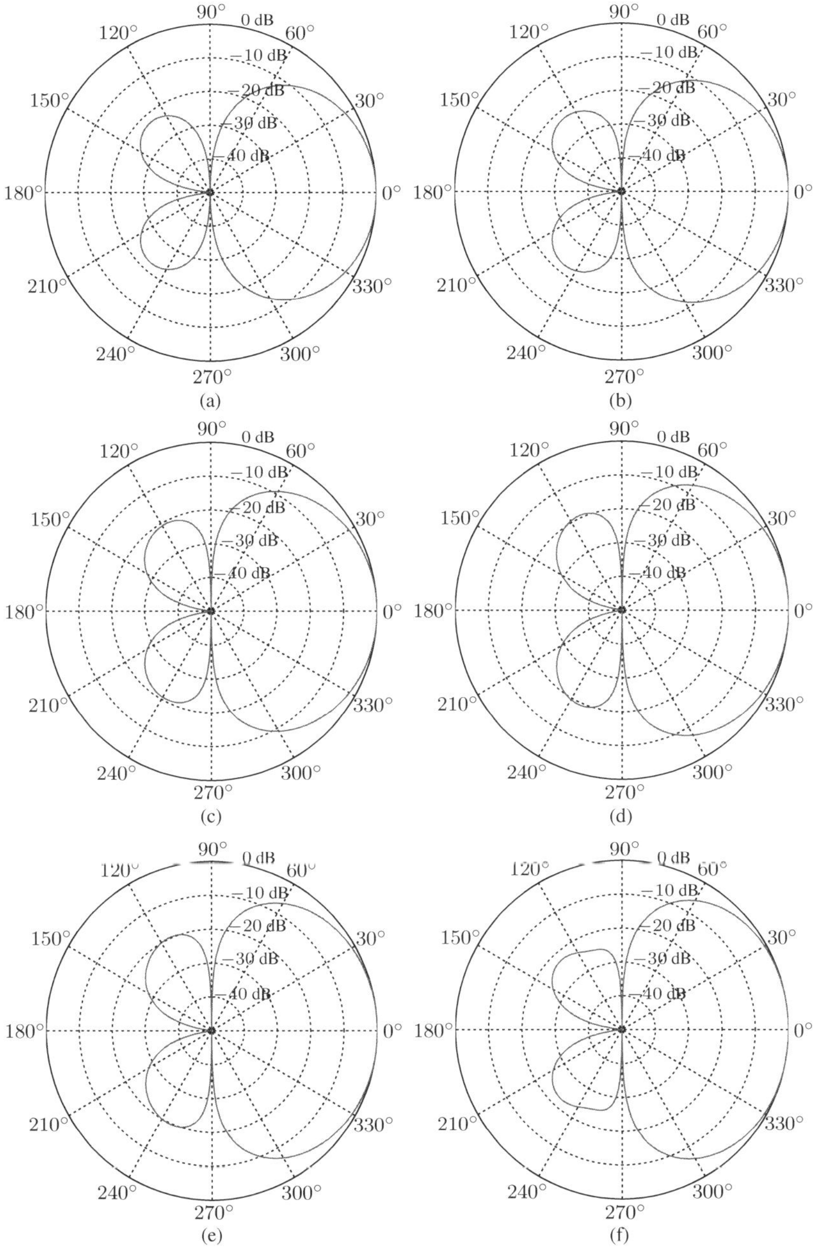

Fig. 6.13 Patterns of the third-order cardioid with 8 microphones for several frequencies: (a) $f = 0.5$ kHz, (b) $f = 1.0$ kHz, (c) $f = 2.0$ kHz, (d) $f = 3.0$ kHz, (e) $f = 4.0$ kHz, and (f) $f = 5.0$ kHz.

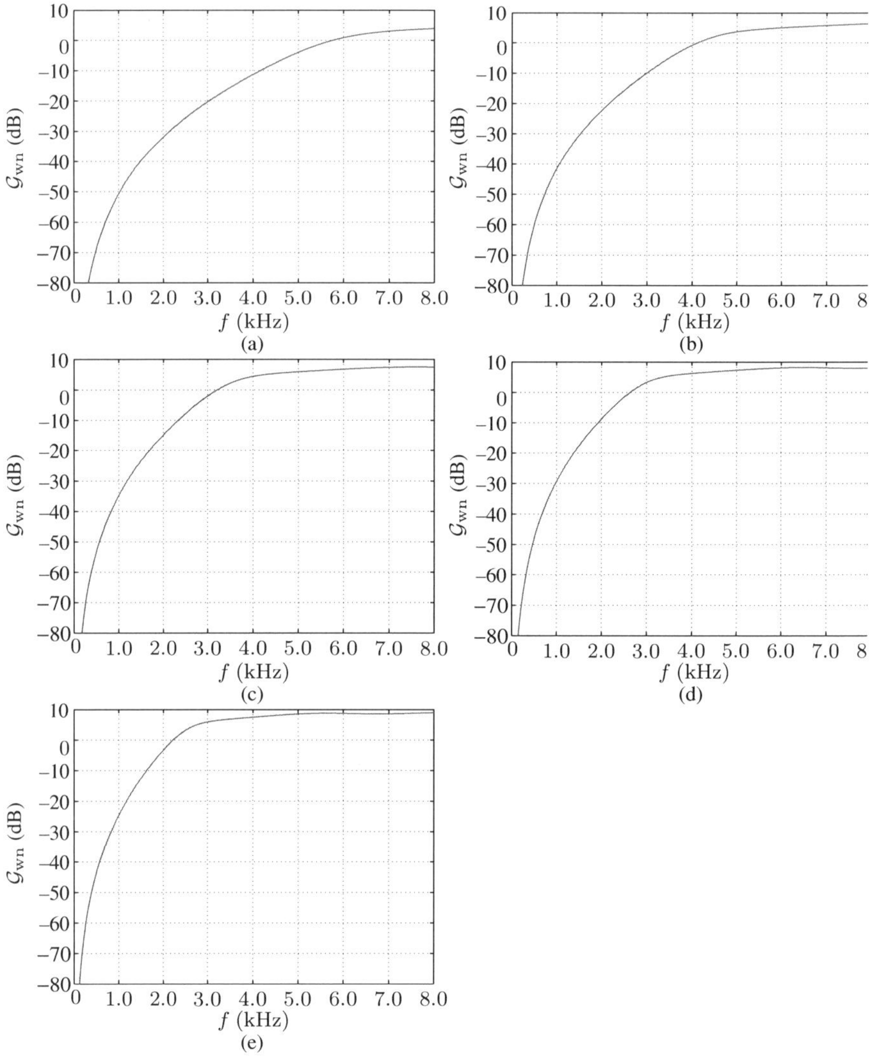

Fig. 6.14 The white noise gain of the third-order cardioid, as a function of frequency, for different values of M: (a) $M = 4$, (b) $M = 5$, (c) $M = 6$, (d) $M = 7$, and (e) $M = 8$.

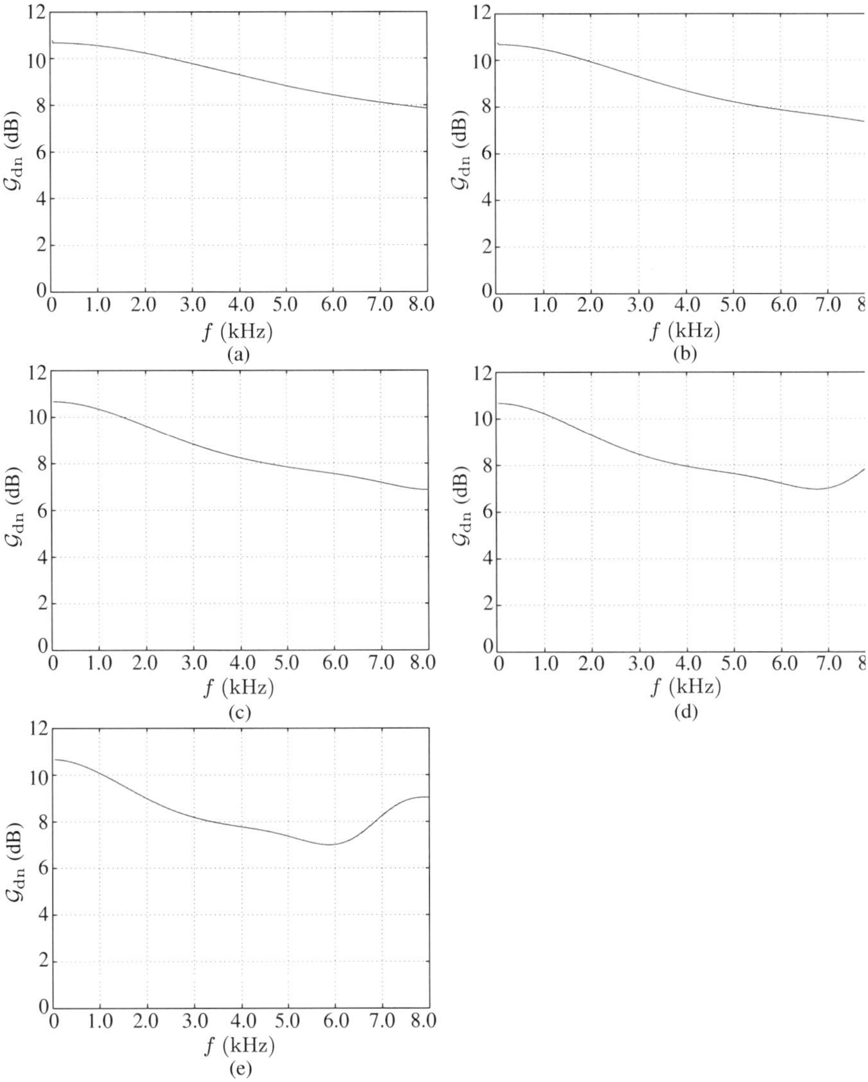

Fig. 6.15 The directivity factor of the third-order cardioid, as a function of frequency, for different values of M: (a) $M = 4$, (b) $M = 5$, (c) $M = 6$, (d) $M = 7$, and (e) $M = 8$.

6.3.4 Third-Order DMA

In this subsection, we propose to design with the minimum-norm filter, the third-order DMA with three distinct nulls proposed in Chapter 5 as Case 1. In this scenario, we have

$$\boldsymbol{\alpha} = \left[1 \; 0 \; -\frac{1}{2} \; -1 \right]^T, \tag{6.27}$$

$$\boldsymbol{\beta} = \left[1 \; 0 \; 0 \; 0 \right]^T. \tag{6.28}$$

Figures 6.16, 6.17, and 6.18 display the patterns of the third-order DMA with three distinct nulls and 4, 5, and 8 microphones, respectively, for several frequencies. All patterns look very similar except for the case $M = 8$ at high frequencies.

In Fig. 6.19, we give plots of the white noise gain of the third-order DMA with three distinct nulls, as a function of frequency, for different values of M. Comparing the 4 and 8 microphone cases at 1 kHz, we observe that the gain is around -48 dB with 4 microphones and around -23 dB with 8 microphones, which represents an improvement of about 25 dB.

In Fig. 6.20, we show plots of the directivity factor of the third-order DMA with three distinct nulls, as a function of frequency, for different values of M. Comparing the 4 and 8 microphone cases, we see that the gain in the second design is only slightly worse in the range 4–7 kHz.

In this chapter, we have shown the effectiveness of the minimum-norm filter in the design of robust DMAs. Fundamentally, we have exploited the fact that we have many more microphones than the order of the DMA. We can further improve the robustness of the DMAs at low frequencies by using the minimum-norm filter with a non-uniform linear array but we leave this study to the reader.

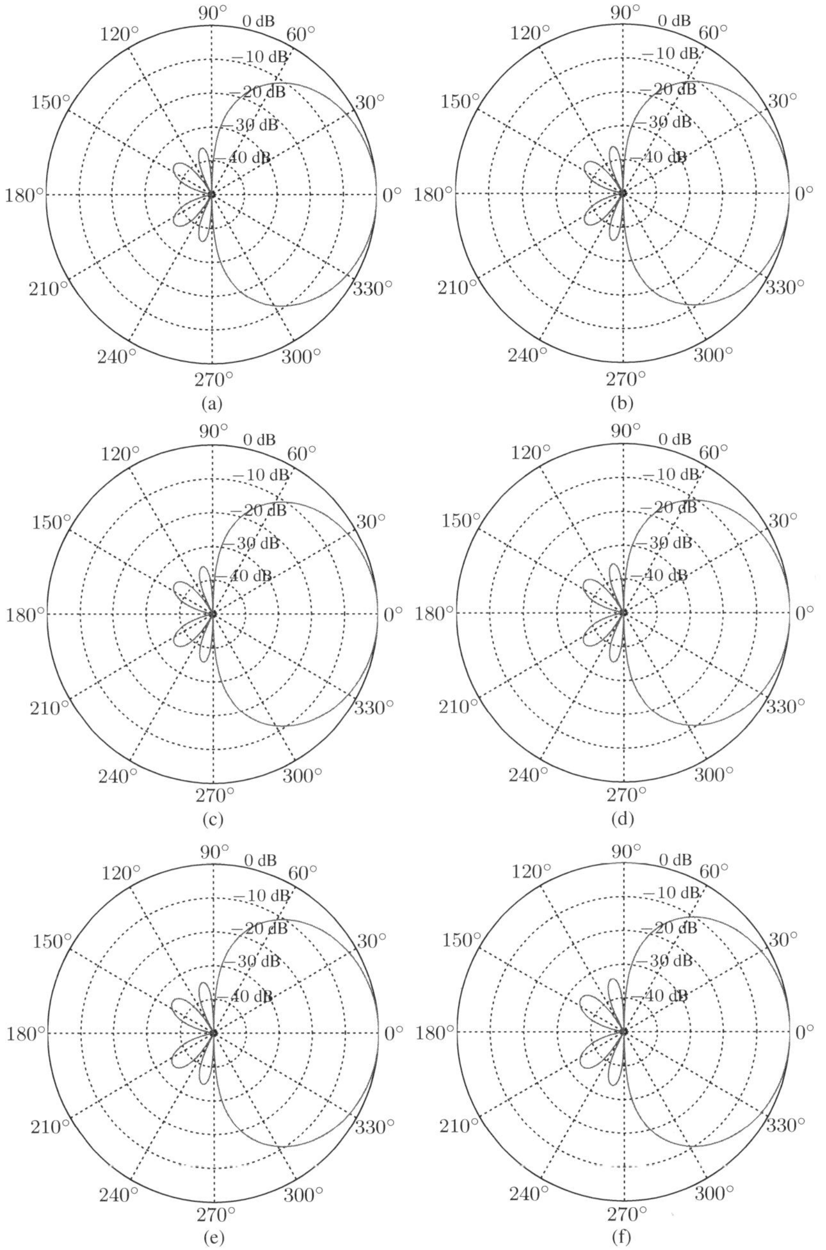

Fig. 6.16 Patterns of the third-order DMA with three distinct nulls and 4 microphones for several frequencies: (a) $f = 0.5$ kHz, (b) $f = 1.0$ kHz, (c) $f = 2.0$ kHz, (d) $f = 3.0$ kHz, (e) $f = 4.0$ kHz, and (f) $f = 5.0$ kHz.

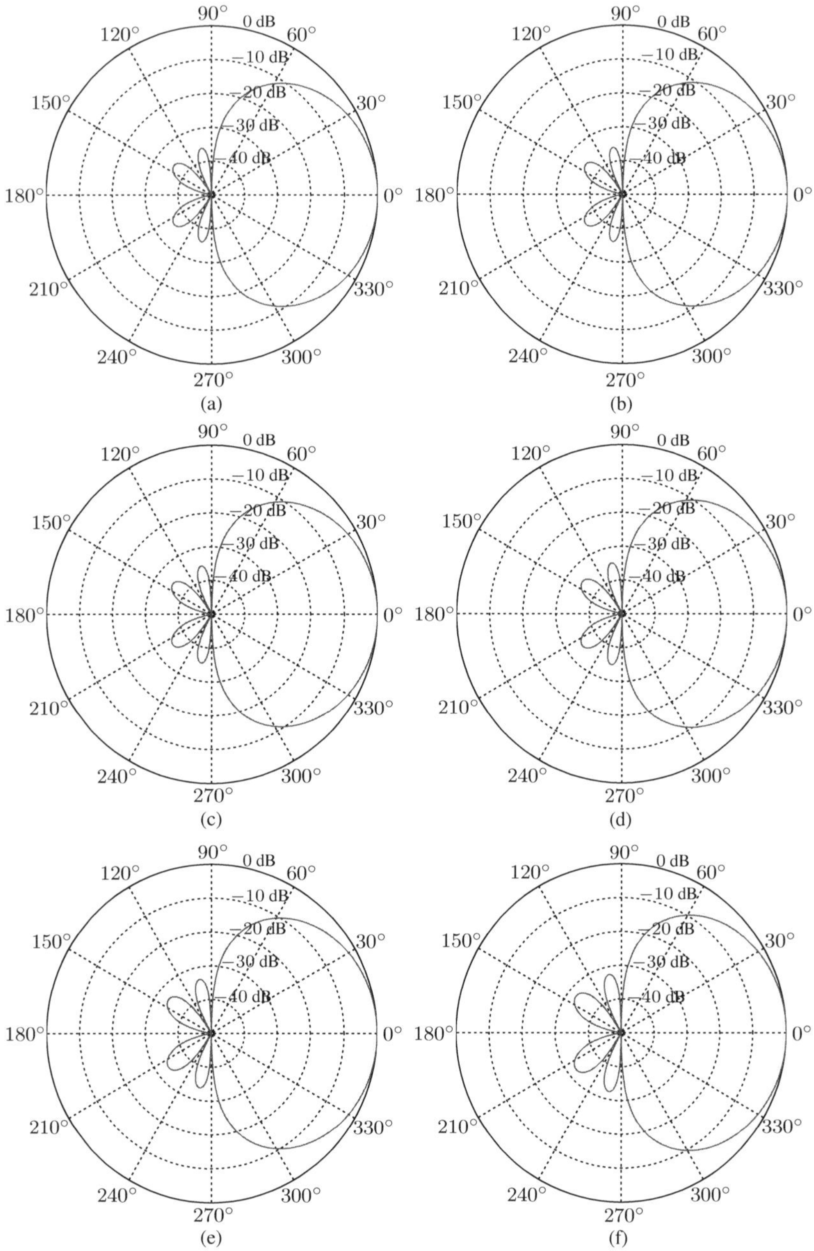

Fig. 6.17 Patterns of the third-order DMA with three distinct nulls and 5 microphones for several frequencies: (a) $f = 0.5$ kHz, (b) $f = 1.0$ kHz, (c) $f = 2.0$ kHz, (d) $f = 3.0$ kHz, (e) $f = 4.0$ kHz, and (f) $f = 5.0$ kHz.

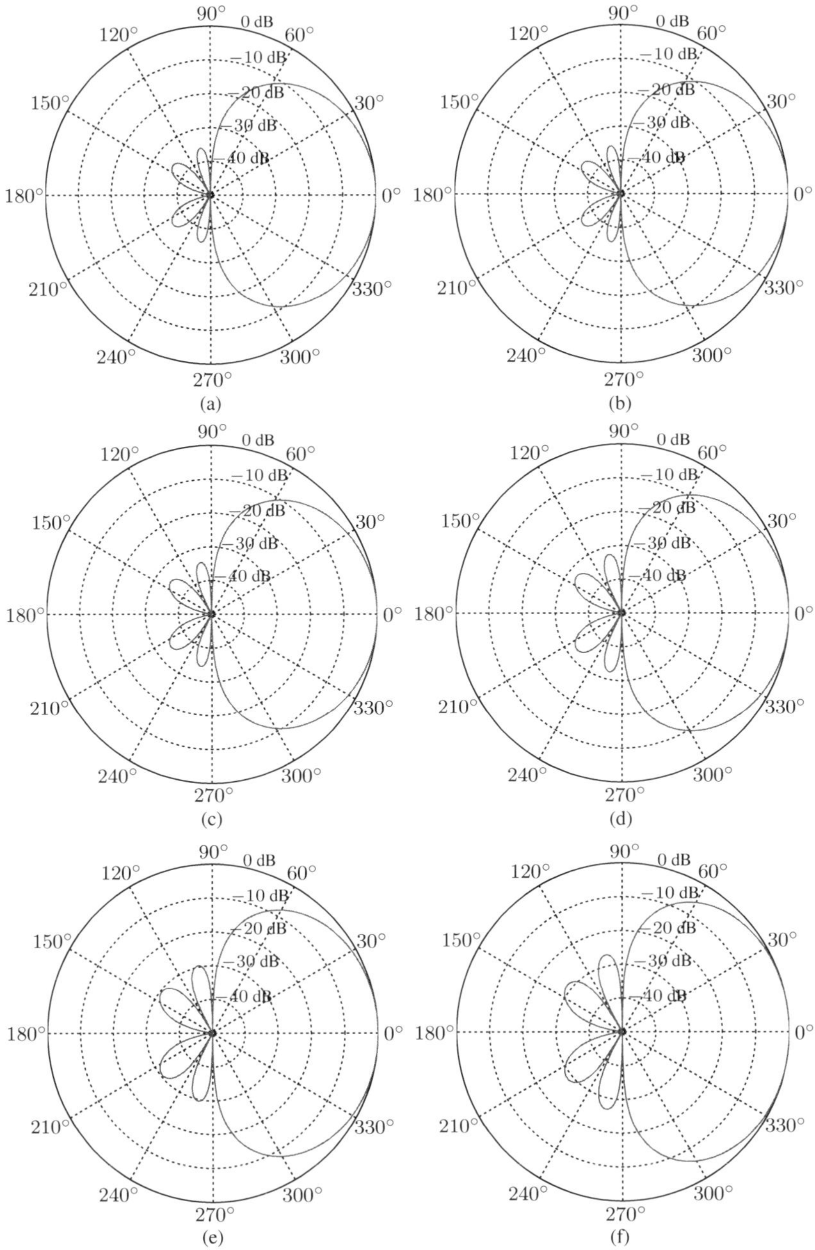

Fig. 6.18 Patterns of the third-order DMA with three distinct nulls and 8 microphones for several frequencies: (a) $f = 0.5$ kHz, (b) $f = 1.0$ kHz, (c) $f = 2.0$ kHz, (d) $f = 3.0$ kHz, (e) $f = 4.0$ kHz, and (f) $f = 5.0$ kHz.

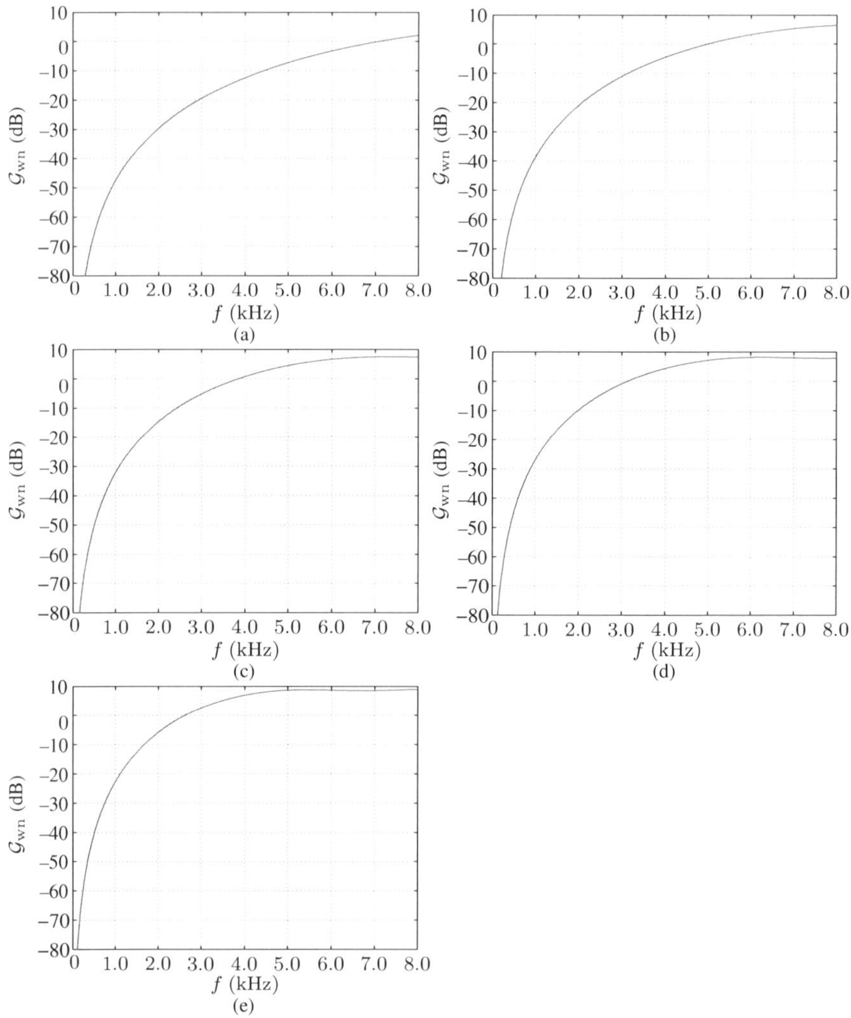

Fig. 6.19 The white noise gain of the third-order DMA with three distinct nulls, as a function of frequency, for different values of M: (a) $M = 4$, (b) $M = 5$, (c) $M = 6$, (d) $M = 7$, and (e) $M = 8$.

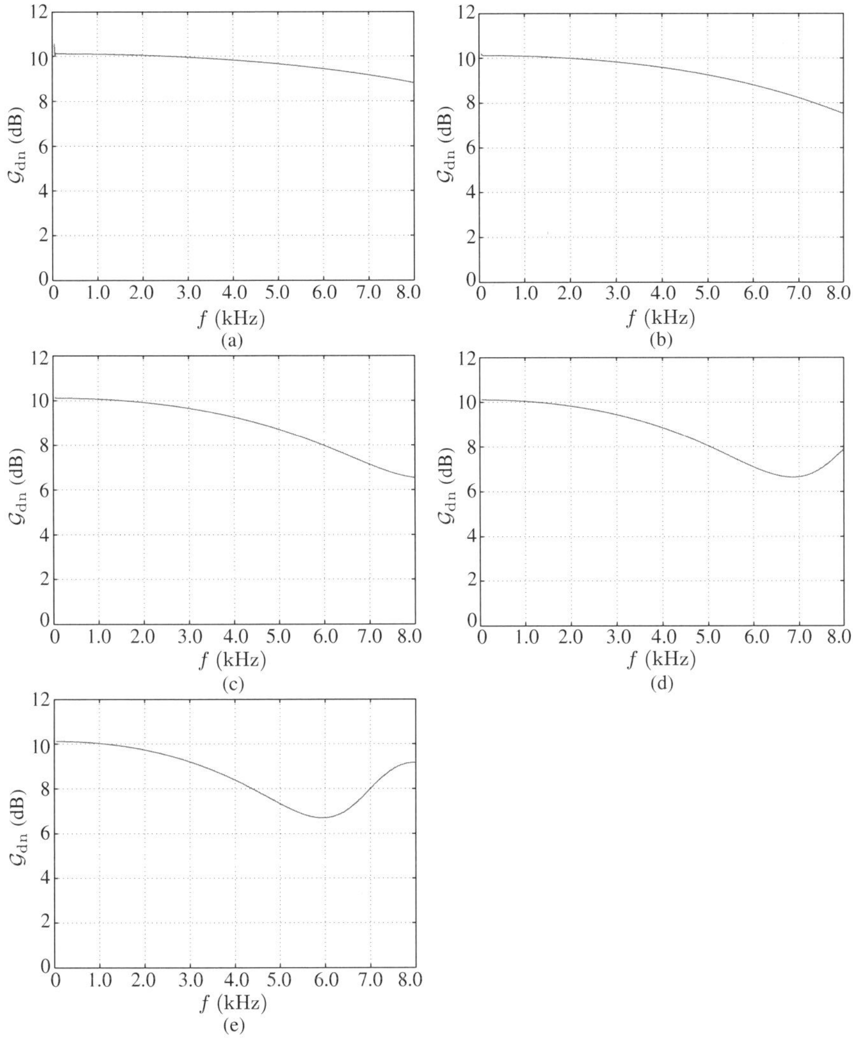

Fig. 6.20 The directivity factor of the third-order DMA with three distinct nulls, as a function of frequency, for different values of M: (a) $M = 4$, (b) $M = 5$, (c) $M = 6$, (d) $M = 7$, and (e) $M = 8$.

References

1. J. Benesty, J. Chen, and Y. Huang, *Microphone Array Signal Processing*. Berlin, Germany: Springer-Verlag, 2008.
2. G. W. Elko, "Superdirectional microphone arrays," in *Acoustic Signal Processing for Telecommunication*, S. L. Gay and J. Benesty, Eds. Boston, MA: Kluwer Academic Publishers, 2000, Chapter 10, pp. 181–237.
3. A. Booker and C. Y. Ong, "Multiple constraint adaptive filtering," *Geophysics*, vol. 36, pp. 498–509, June 1971.
4. O. Frost, "An algorithm for linearly constrained adaptive array processing," *Proc. IEEE*, vol. 60, pp. 926–935, Jan. 1972.
5. H. Cox, R. M. Zeskind, and M. M. Owen, "Robust adaptive beamforming," *IEEE Trans. Acoust., Speech, Signal Process.*, vol. ASSP-35, pp. 1365–1376, Oct. 1987.
6. J. Benesty and J. Chen, *Optimal Noise Reduction Filters–A Theoretical Study.* Springer Briefs in Electrical & Computer Engineering, 2011.
7. J. Capon, "High resolution frequency-wavenumber spectrum analysis," *Proc. IEEE*, vol. 57, pp. 1408–1418, Aug. 1969.
8. R. T. Lacoss, "Data adaptive spectral analysis methods," *Geophysics*, vol. 36, pp. 661–675, Aug. 1971.
9. J. Benesty, J. Chen, and E. Habets, *Speech Enhancement in the STFT Domain.* Springer Briefs in Electrical & Computer Engineering, 2011.

Chapter 7
Study and Design of Differential Arrays with the MacLaurin's Series Approximation

In Chapter 2, we showed that the patterns of differential arrays can be obtained from the general definition of the beampattern by approximating the exponential function with its MacLaurin's series expansion. In other words, a directional pattern of order N can be obtained from the MacLaurin's series of order N, as long as this approximation holds. In this chapter, we show how to design differential arrays based on this approach and their relationship to adaptive beamforming. This investigation is far from complete and more can be done; our aim here is just to show the potential of this new concept.

7.1 Principle

In Chapter 2, we demonstrated that there is an interesting relationship between the coefficients of a directional pattern of order N and the coefficients of a beamforming filter of length $N+1$, i.e.,

$$a_{N,n} = \frac{\left(\jmath\omega\tau_0\right)^n}{n!} \sum_{m=1}^{N+1} (m-1)^n H_m(\omega), \ n = 0, 1, \ldots, N. \tag{7.1}$$

In the previous expression, it is assumed that given a uniform linear array of $N+1$ microphones, we have

$$e^x - \sum_{n=0}^{N} \frac{x^n}{n!}, \tag{7.2}$$

meaning that the variable x, which depends on the sensor spacing, is relatively small [1].

From (7.1), it is easy to deduce the linear system of $N+1$ equations:

$$\boldsymbol{\Psi}_{N+1}\mathbf{h}(\omega) = \mathbf{a}_{N+1}(\omega), \tag{7.3}$$

J. Benesty and J. Chen, *Study and Design of Differential Microphone Arrays*, Springer Topics in Signal Processing 6, DOI: 10.1007/978-3-642-33753-6_7,

where

$$\mathbf{\Psi}_{N+1} = \begin{bmatrix} 1 & 1 & 1 & \cdots & 1 \\ 0 & 1 & 2 & \cdots & N \\ 0 & 1 & 2^2 & \cdots & N^2 \\ \vdots & \vdots & \vdots & \ddots & \vdots \\ 0 & 1 & 2^N & \cdots & N^N \end{bmatrix} \tag{7.4}$$

is a Vandermonde matrix of size $(N+1) \times (N+1)$ and

$$\mathbf{a}_{N+1}(\omega) = \begin{bmatrix} a_{N,0} & \frac{a_{N,1}}{\jmath\omega\tau_0} & \frac{2!\, a_{N,2}}{(\jmath\omega\tau_0)^2} & \cdots & \frac{N!\, a_{N,N}}{(\jmath\omega\tau_0)^N} \end{bmatrix}^T \tag{7.5}$$

is a vector of length $N+1$.

In the next sections, we will study and design different order differential arrays based on this approach.

7.2 First-Order Differential Arrays

The system of two linear equations corresponding to first-order DMAs is

$$\begin{bmatrix} 1 & 1 \\ 0 & 1 \end{bmatrix} \mathbf{h}(\omega) = \begin{bmatrix} a_{1,0} \\ \frac{a_{1,1}}{\jmath\omega\tau_0} \end{bmatrix}, \tag{7.6}$$

for which the solution is

$$\mathbf{h}(\omega) = \frac{\jmath a_{1,1}}{\omega\tau_0} \begin{bmatrix} 1 - \jmath\omega\tau_0 \frac{a_{1,0}}{a_{1,1}} \\ -1 \end{bmatrix}. \tag{7.7}$$

Replacing $1+x$ by e^x in (7.7), we get

$$\mathbf{h}(\omega) = \frac{\jmath a_{1,1}}{\omega\tau_0} \begin{bmatrix} e^{-\jmath\omega\tau_0 a_{1,0}/a_{1,1}} \\ -1 \end{bmatrix}. \tag{7.8}$$

We then deduce that the equivalent filter is

$$\mathbf{h}'(\omega) = \frac{1}{\omega} \begin{bmatrix} -e^{-\jmath\omega\tau_0 a_{1,0}/a_{1,1}} \\ 1 \end{bmatrix}, \tag{7.9}$$

since $\mathcal{C}_1 = -\frac{\jmath a_{1,1}}{\tau_0}$ is frequency independent. In the rest of this section, we will consider the filter $\mathbf{h}'(\omega)$ as the practical beamformer that should be implemented no matter if (7.2) holds or not, while the filter $\mathbf{h}(\omega)$ is considered as the ideal filter for which (7.2) holds.

The beampattern of the beamformer $\mathbf{h}'(\omega)$ is

$$\begin{aligned}\mathcal{B}\left[\mathbf{h}'(\omega),\theta\right] &= \mathbf{d}^H(\omega,\cos\theta)\,\mathbf{h}'(\omega)\\ &= \frac{1}{\omega}\left(-e^{-\jmath\omega\tau_0 a_{1,0}/a_{1,1}} + e^{\jmath\omega\tau_0\cos\theta}\right),\end{aligned} \tag{7.10}$$

therefore,

$$\left|\mathcal{B}\left[\mathbf{h}'(\omega),\theta\right]\right|^2 = \frac{2}{\omega^2}\left[1-\cos\left(\frac{\omega\tau_0}{a_{1,1}}\right)\right], \tag{7.11}$$

and taking

$$\frac{1}{a_{1,1}} = 1-\alpha_{1,1} \tag{7.12}$$

in (7.11), we obtain the exact same patterns as the ones obtained in Chapter 3 for the dipole, cardioid, hypercardioid, and supercardioid.

It is of interest to verify that, indeed, when (7.2) holds, we have

$$\mathcal{B}\left[\mathbf{h}(\omega),\theta\right] = a_{1,0}+a_{1,1}\cos\theta. \tag{7.13}$$

The white noise gain of the beamformer $\mathbf{h}'(\omega)$ is

$$\begin{aligned}\mathcal{G}_{\mathrm{wn},1}\left[\mathbf{h}'(\omega)\right] &= \frac{\left|\mathbf{h}'^H(\omega)\,\mathbf{d}(\omega,1)\right|^2}{\mathbf{h}'^H(\omega)\,\mathbf{h}'(\omega)}\\ &= \frac{1}{2}\left|e^{-\jmath\omega\tau_0 a_{1,0}/a_{1,1}} - e^{\jmath\omega\tau_0}\right|^2\\ &= 1-\cos\left(\frac{\omega\tau_0}{a_{1,1}}\right).\end{aligned} \tag{7.14}$$

Taking again (7.12) in (7.14), we obtain the exact same figures as the ones obtained in Chapter 3 for the dipole, cardioid, hypercardioid, and supercardioid.

For the beamformer $\mathbf{h}(\omega)$, the white noise gain is

$$\begin{aligned}\mathcal{G}_{\mathrm{wn},1}\left[\mathbf{h}(\omega)\right] &= \frac{\left|\mathbf{h}^H(\omega)\,\mathbf{d}(\omega,1)\right|^2}{\mathbf{h}^H(\omega)\,\mathbf{h}(\omega)}\\ &= \frac{(\omega\tau_0)^2}{2a_{1,1}^2+(\omega\tau_0)^2\,a_{1,0}^2}.\end{aligned} \tag{7.15}$$

We conclude that the white noise is amplified if

$$\omega\tau_0 < \sqrt{\frac{2a_{1,1}}{2-a_{1,1}}}. \tag{7.16}$$

The directivity factor of the beamformer $\mathbf{h}'(\omega)$ is

$$\begin{aligned}\mathcal{G}_{\mathrm{dn},1}\left[\mathbf{h}'(\omega)\right] &= \frac{\left|\mathbf{h}'^H(\omega)\,\mathbf{d}(\omega,1)\right|^2}{\mathbf{h}'^H(\omega)\,\mathbf{\Gamma}_{\mathrm{dn}}(\omega)\,\mathbf{h}'(\omega)} \\ &= \frac{1-\cos\left(\dfrac{\omega\tau_0}{a_{1,1}}\right)}{1-\mathrm{sinc}(\omega\tau_0)\cos\left(\dfrac{\omega\tau_0 a_{1,0}}{a_{1,1}}\right)}.\end{aligned} \tag{7.17}$$

Again, using the relation (7.12) in (7.17), we easily find that

$$\mathcal{G}_{\mathrm{dn},1}\left[\mathbf{h}'(\omega)\right] = \frac{1-\cos\left[\omega\tau_0\left(1-\alpha_{1,1}\right)\right]}{1-\mathrm{sinc}(\omega\tau_0)\cos\left(\omega\tau_0\alpha_{1,1}\right)}. \tag{7.18}$$

This gain is identical to the one derived in Chapter 3 [eq. (3.51)].

For the beamformer $\mathbf{h}(\omega)$, the directivity factor is

$$\begin{aligned}\mathcal{G}_{\mathrm{dn},1}\left[\mathbf{h}(\omega)\right] &= \frac{\left|\mathbf{h}^H(\omega)\,\mathbf{d}(\omega,1)\right|^2}{\mathbf{h}^H(\omega)\,\mathbf{\Gamma}_{\mathrm{dn}}(\omega)\,\mathbf{h}(\omega)} \\ &= \frac{1}{\frac{4}{3}a_{1,1}^2-2a_{1,1}+1}.\end{aligned} \tag{7.19}$$

Substituting (7.12) into (7.19), we get

$$\mathcal{G}_{\mathrm{dn},1}\left[\mathbf{h}(\omega)\right] = 3\frac{\left(1-\alpha_{1,1}\right)^2}{3\alpha_{1,1}^2+1}, \tag{7.20}$$

which is identical to the formula obtained in Chapter 3 [eq. (3.52)].

The gain of $\mathbf{h}'(\omega)$ for a point noise source is

$$\begin{aligned}\mathcal{G}_{\mathrm{ns},1}\left[\mathbf{h}'(\omega)\right] &= \frac{\left|\mathbf{h}'^H(\omega)\,\mathbf{d}(\omega,1)\right|^2}{\left|\mathbf{h}'^H(\omega)\,\mathbf{d}(\omega,\cos\theta_{\mathrm{n}})\right|^2} \\ &= \frac{\left|e^{-\jmath\omega\tau_0 a_{1,0}/a_{1,1}}-e^{\jmath\omega\tau_0}\right|^2}{\left|e^{-\jmath\omega\tau_0 a_{1,0}/a_{1,1}}-e^{\jmath\omega\tau_0\cos\theta_{\mathrm{n}}}\right|^2} \\ &= \frac{1-\cos\left(\dfrac{\omega\tau_0}{a_{1,1}}\right)}{1-\cos\left[\omega\tau_0\left(\cos\theta_{\mathrm{n}}+\dfrac{a_{1,0}}{a_{1,1}}\right)\right]}\end{aligned} \tag{7.21}$$

and using (7.12), we obtain the gain derived in Chapter 3 [eq. (3.55)], i.e.,

$$\mathcal{G}_{\text{ns},1}\left[\mathbf{h}'(\omega)\right] = \frac{1-\cos\left[\omega\tau_0\left(1-\alpha_{1,1}\right)\right]}{1-\cos\left[\omega\tau_0\left(\cos\theta_{\text{n}}-\alpha_{1,1}\right)\right]}. \tag{7.22}$$

The gain of $\mathbf{h}(\omega)$ for a point noise source is

$$\begin{aligned}\mathcal{G}_{\text{ns},1}\left[\mathbf{h}(\omega)\right] &= \frac{\left|\mathbf{h}^H(\omega)\,\mathbf{d}(\omega,1)\right|^2}{\left|\mathbf{h}^H(\omega)\,\mathbf{d}(\omega,\cos\theta_{\text{n}})\right|^2} \\ &= \frac{1}{\left(a_{1,0}+a_{1,1}\cos\theta_{\text{n}}\right)^2},\end{aligned} \tag{7.23}$$

which corresponds to the theoretical gain of the first-order DMA.

Clearly, the first-order DMAs derived in this section are strictly equivalent to the ones derived in Chapter 3.

7.3 Second-Order Differential Arrays

For second-order DMAs, we need to inverse the matrix $\mathbf{\Psi}_3$, which is given by

$$\mathbf{\Psi}_3^{-1} = \begin{bmatrix} 1 & -\frac{3}{2} & \frac{1}{2} \\ 0 & 2 & -1 \\ 0 & -\frac{1}{2} & \frac{1}{2} \end{bmatrix}. \tag{7.24}$$

We deduce that the corresponding filter is

$$\begin{aligned}\mathbf{h}(\omega) &= \begin{bmatrix} a_{2,0} - \dfrac{a_{2,2}}{(\omega\tau_0)^2} + \jmath\dfrac{3a_{2,1}}{2\omega\tau_0} \\ \dfrac{2a_{2,2}}{(\omega\tau_0)^2} - \jmath\dfrac{2a_{2,1}}{\omega\tau_0} \\ -\dfrac{a_{2,2}}{(\omega\tau_0)^2} + \jmath\dfrac{a_{2,1}}{2\omega\tau_0} \end{bmatrix} \\ &= -\frac{a_{2,2}}{(\omega\tau_0)^2}\begin{bmatrix} 1 - \dfrac{a_{2,0}(\omega\tau_0)^2}{a_{2,2}} - \jmath\dfrac{3\omega\tau_0 a_{2,1}}{2a_{2,2}} \\ -2 + \jmath\dfrac{2\omega\tau_0 a_{2,1}}{a_{2,2}} \\ 1 - \jmath\dfrac{\omega\tau_0 a_{2,1}}{2a_{2,2}} \end{bmatrix}.\end{aligned} \tag{7.25}$$

Replacing $1+x+x^2/2$ by e^x in (7.25), we get

$$\mathbf{h}(\omega) = -\frac{a_{2,2}}{(\omega\tau_0)^2}\begin{bmatrix} e^{-\jmath\omega\tau_0 3a_{2,1}/2a_{2,2}} + \left(\frac{9a_{2,1}^2}{8a_{2,2}^2} - \frac{a_{2,0}}{a_{2,2}}\right)(\omega\tau_0)^2 \\ -2e^{-\jmath\omega\tau_0 a_{2,1}/a_{2,2}} - \frac{a_{2,1}^2}{a_{2,2}^2}(\omega\tau_0)^2 \\ e^{-\jmath\omega\tau_0 a_{2,1}/2a_{2,2}} + \frac{a_{2,1}^2}{8a_{2,2}^2}(\omega\tau_0)^2 \end{bmatrix}. \quad (7.26)$$

Since $\mathcal{C}_2 = -\frac{a_{2,2}}{\tau_0^2}$ is constant across frequencies, the equivalent filter is

$$\mathbf{h}'(\omega) = \frac{1}{\omega^2}\begin{bmatrix} e^{-\jmath\omega\tau_0 3a_{2,1}/2a_{2,2}} + \left(\frac{9a_{2,1}^2}{8a_{2,2}^2} - \frac{a_{2,0}}{a_{2,2}}\right)(\omega\tau_0)^2 \\ -2e^{-\jmath\omega\tau_0 a_{2,1}/a_{2,2}} - \frac{a_{2,1}^2}{a_{2,2}^2}(\omega\tau_0)^2 \\ e^{-\jmath\omega\tau_0 a_{2,1}/2a_{2,2}} + \frac{a_{2,1}^2}{8a_{2,2}^2}(\omega\tau_0)^2 \end{bmatrix}. \quad (7.27)$$

As we did it in the previous section, we consider $\mathbf{h}(\omega)$ as the ideal filter for which (7.2) holds and $\mathbf{h}'(\omega)$ the practical beamformer that should be implemented no matter if (7.2) holds or not.

It can be verified that

$$\begin{aligned} \mathcal{B}[\mathbf{h}(\omega), \theta] &= \mathbf{d}^H(\omega, \cos\theta)\,\mathbf{h}(\omega) \\ &= \sum_{n=1}^{3} H_n(\omega)\, e^{\jmath\omega(n-1)\tau_0\cos\theta} \\ &= H_1(\omega) + H_2(\omega)\left[1 + \jmath\omega\tau_0\cos\theta - (\omega\tau_0)^2\frac{\cos^2\theta}{2}\right] \\ &\quad + H_3(\omega)\left[1 + 2\jmath\omega\tau_0\cos\theta - 2(\omega\tau_0)^2\cos^2\theta\right] \\ &= a_{2,0} + a_{2,1}\cos\theta + a_{2,2}\cos^2\theta, \end{aligned} \quad (7.28)$$

while

$$\mathcal{B}[\mathbf{h}'(\omega), \theta] = \sum_{n=1}^{3} H'_n(\omega)\, e^{\jmath\omega(n-1)\tau_0\cos\theta}. \quad (7.29)$$

Figures 7.1, 7.2, 7.3, and 7.4 display the patterns from (7.29) for the dipole, cardioid, hypercardioid, and supercardioid, respectively, for several frequencies and two values of δ.

The white noise gain of the beamformer $\mathbf{h}'(\omega)$ is

$$\mathcal{G}_{\text{wn},2}[\mathbf{h}'(\omega)] = \frac{\left|\sum_{n=1}^{3} H'_n(\omega)\, e^{\jmath\omega(n-1)\tau_0}\right|^2}{\sum_{n=1}^{3}|H'_n(\omega)|^2}. \quad (7.30)$$

In Figs. 7.5, 7.6, 7.7, and 7.8, we plot $\mathcal{G}_{\text{wn},2}\left[\mathbf{h}'(\omega)\right]$ for the dipole, cardioid, hypercardioid, and supercardioid, respectively, for several frequencies and different values of δ.

For the beamformer $\mathbf{h}'(\omega)$, the directivity factor is

$$\mathcal{G}_{\text{dn},2}\left[\mathbf{h}'(\omega)\right] = \frac{\left|\sum_{n=1}^{3} H_n'(\omega)\, e^{\jmath\omega(n-1)\tau_0}\right|^2}{\mathbf{h}'^H(\omega)\, \mathbf{\Gamma}_{\text{dn}}(\omega)\, \mathbf{h}'(\omega)}. \tag{7.31}$$

In Figs 7.9, 7.10, 7.11, and 7.12, we plot $\mathcal{G}_{\text{dn},2}\left[\mathbf{h}'(\omega)\right]$ for the dipole, cardioid, hypercardioid, and supercardioid, respectively, for several frequencies and different values of δ.

The gain of $\mathbf{h}'(\omega)$ for a point noise source is

$$\mathcal{G}_{\text{ns},2}\left[\mathbf{h}'(\omega)\right] = \frac{\left|\sum_{n=1}^{3} H_n'(\omega)\, e^{\jmath\omega(n-1)\tau_0}\right|^2}{\left|\sum_{n=1}^{3} H_n'(\omega)\, e^{\jmath\omega(n-1)\tau_0\cos\theta_{\text{n}}}\right|^2}. \tag{7.32}$$

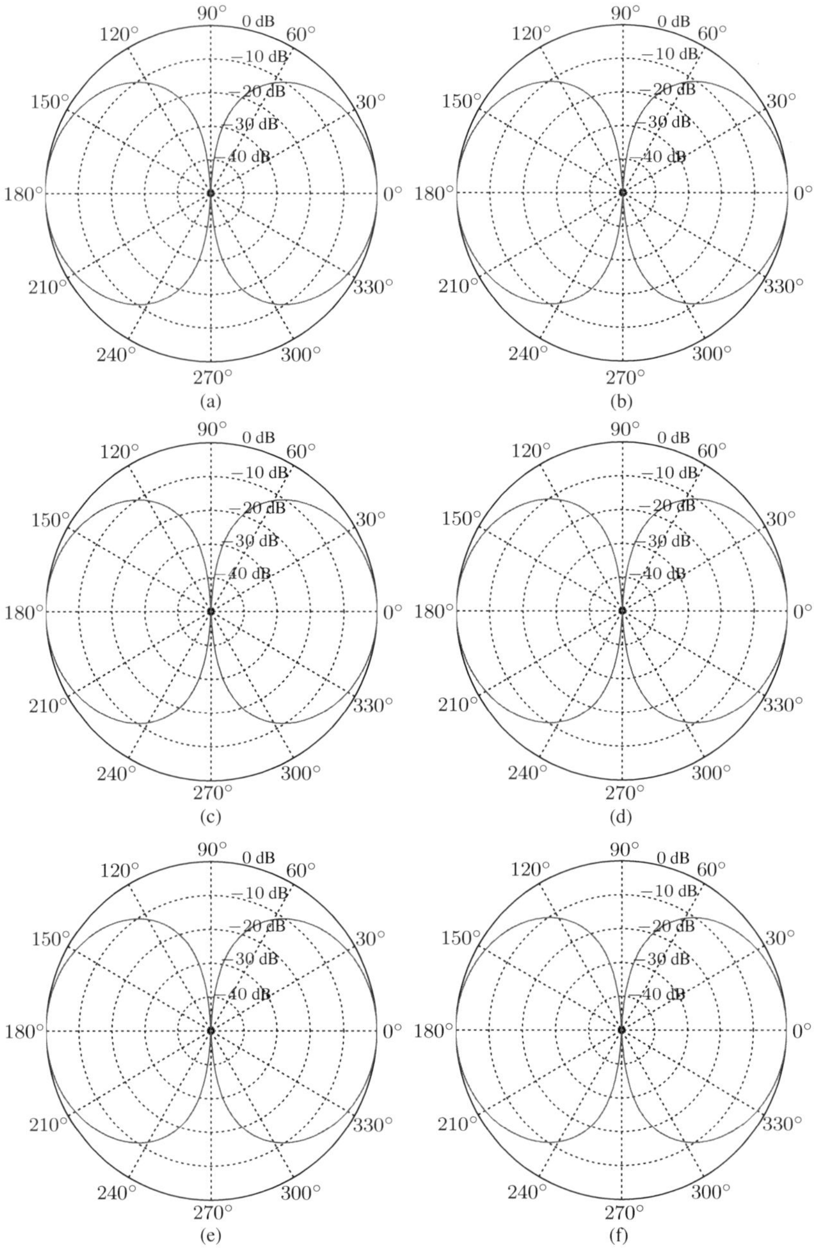

Fig. 7.1 Patterns of the second-order dipole for several frequencies and two values of δ: (a) $f = 0.5$ kHz, $\delta = 0.5$ cm, (b) $f = 0.5$ kHz, $\delta = 1$ cm, (c) $f = 1$ kHz, $\delta = 0.5$ cm, (d) $f = 1$ kHz, $\delta = 1$ cm, (e) $f = 3$ kHz, $\delta = 0.5$ cm, and (f) $f = 3$ kHz, $\delta = 1$ cm.

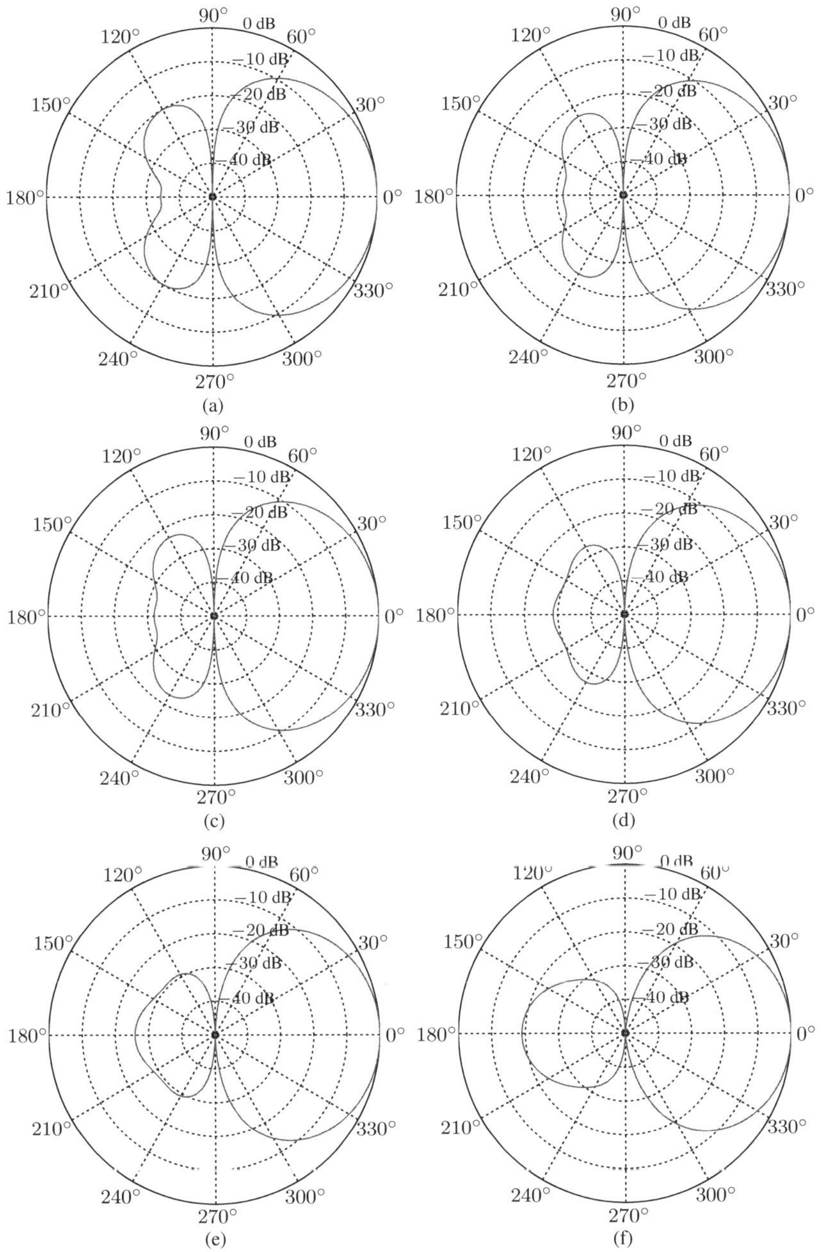

Fig. 7.2 Patterns of the second-order cardioid for several frequencies and two values of δ: (a) $f = 0.5$ kHz, $\delta = 0.5$ cm, (b) $f = 0.5$ kHz, $\delta = 1$ cm, (c) $f = 1$ kHz, $\delta = 0.5$ cm, (d) $f = 1$ kHz, $\delta = 1$ cm, (e) $f = 3$ kHz, $\delta = 0.5$ cm, and (f) $f = 3$ kHz, $\delta = 1$ cm.

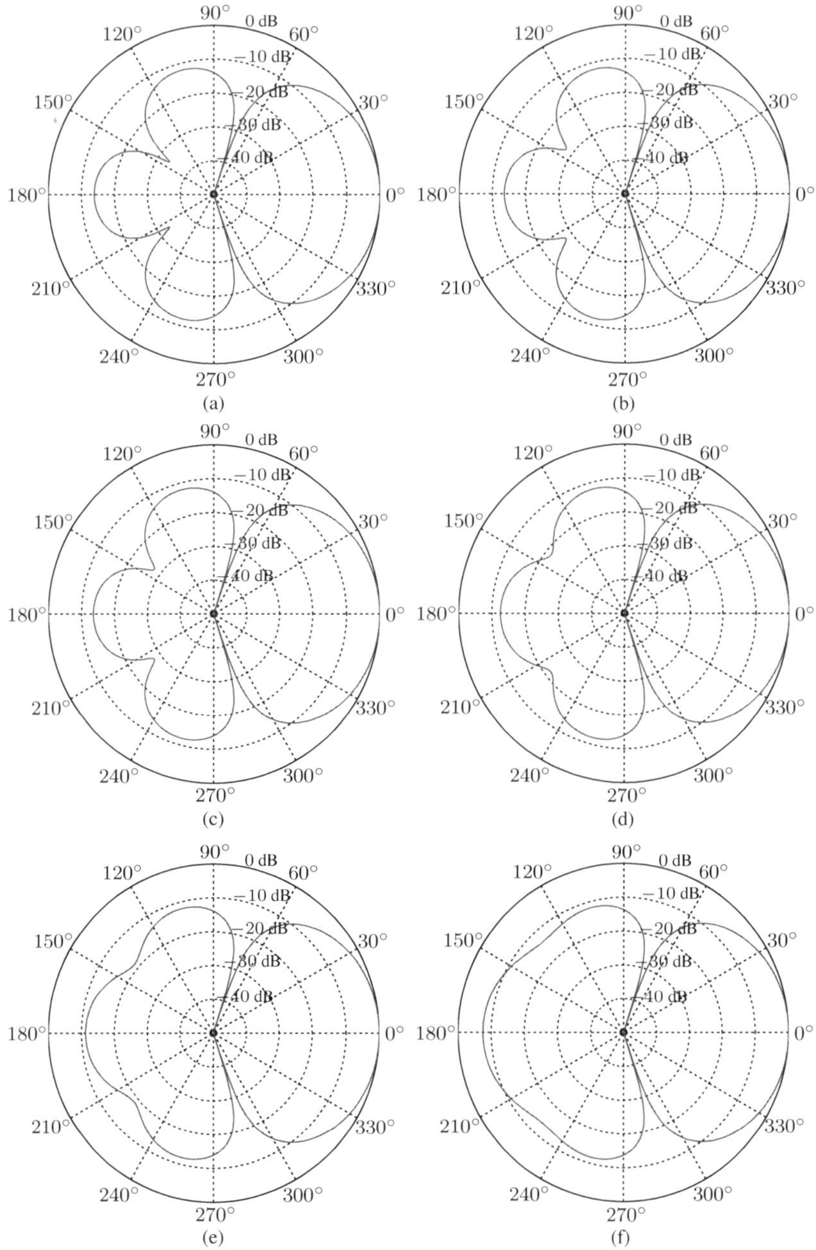

Fig. 7.3 Patterns of the second-order hypercardioid for several frequencies and two values of δ: (a) $f = 0.5$ kHz, $\delta = 0.5$ cm, (b) $f = 0.5$ kHz, $\delta = 1$ cm, (c) $f = 1$ kHz, $\delta = 0.5$ cm, (d) $f = 1$ kHz, $\delta = 1$ cm, (e) $f = 3$ kHz, $\delta = 0.5$ cm, and (f) $f = 3$ kHz, $\delta = 1$ cm.

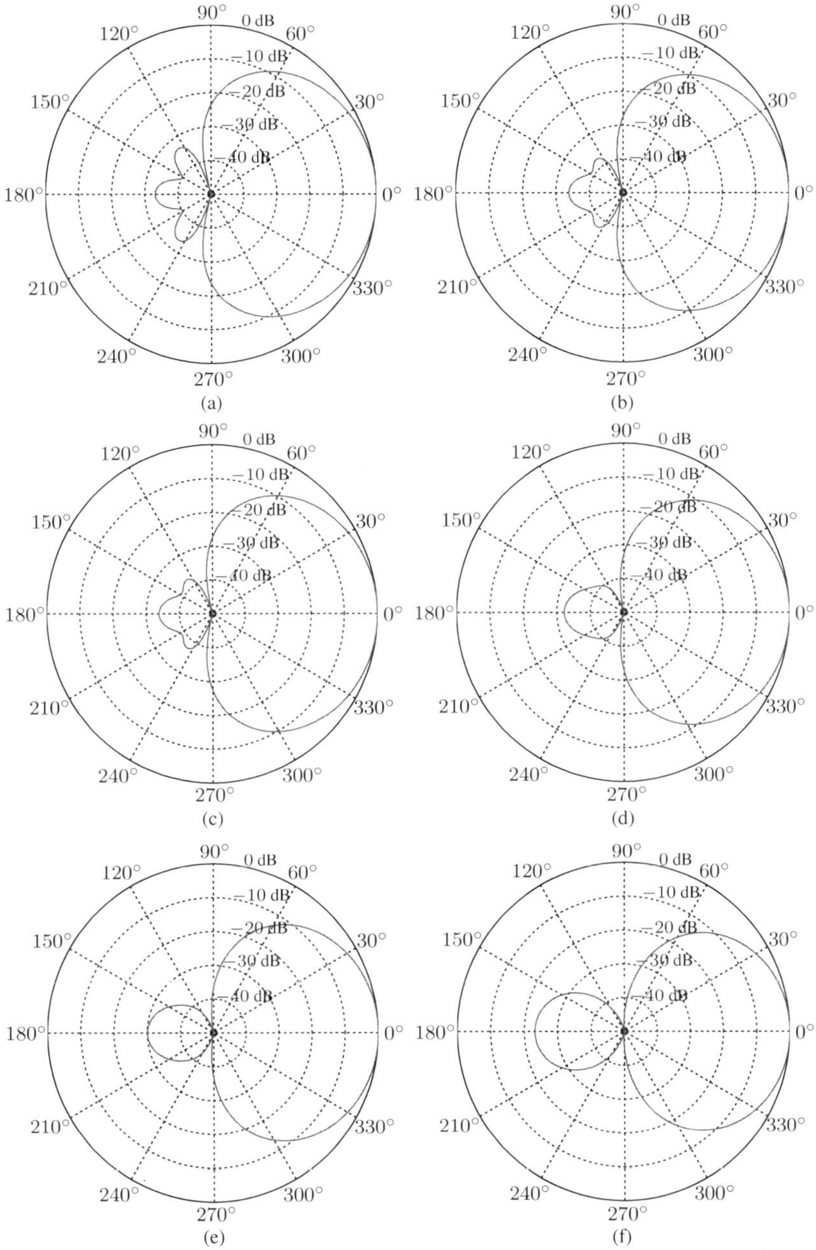

Fig. 7.4 Patterns of the second-order supercardioid for several frequencies and two values of δ: (a) $f = 0.5$ kHz, $\delta = 0.5$ cm, (b) $f = 0.5$ kHz, $\delta = 1$ cm, (c) $f = 1$ kHz, $\delta = 0.5$ cm, (d) $f = 1$ kHz, $\delta = 1$ cm, (e) $f = 3$ kHz, $\delta = 0.5$ cm, and (f) $f = 3$ kHz, $\delta = 1$ cm.

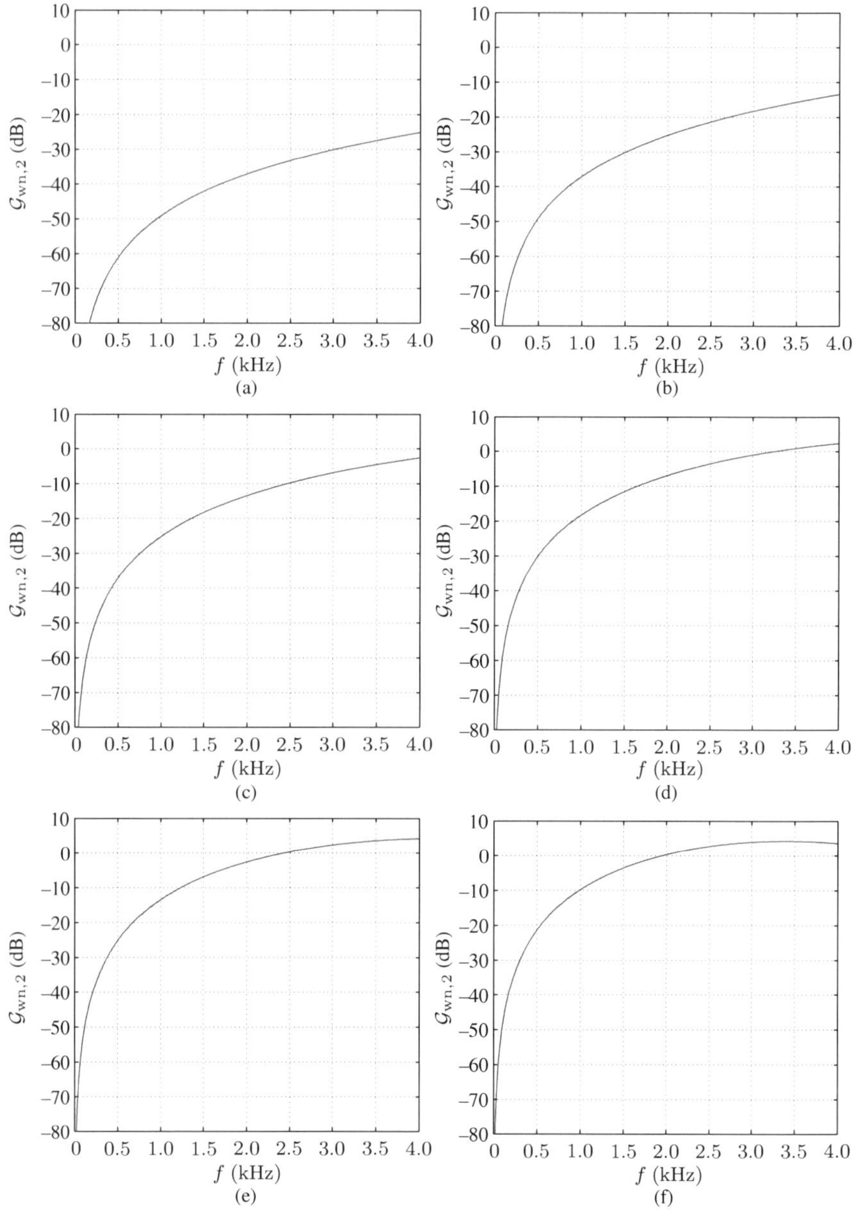

Fig. 7.5 The white noise gain of the second-order dipole, as a function of frequency, for different values of δ: (a) $\delta = 0.5$ cm, (b) $\delta = 1$ cm, (c) $\delta = 2$ cm, (d) $\delta = 3$ cm, (e) $\delta = 4$ cm, and (f) $\delta = 5$ cm.

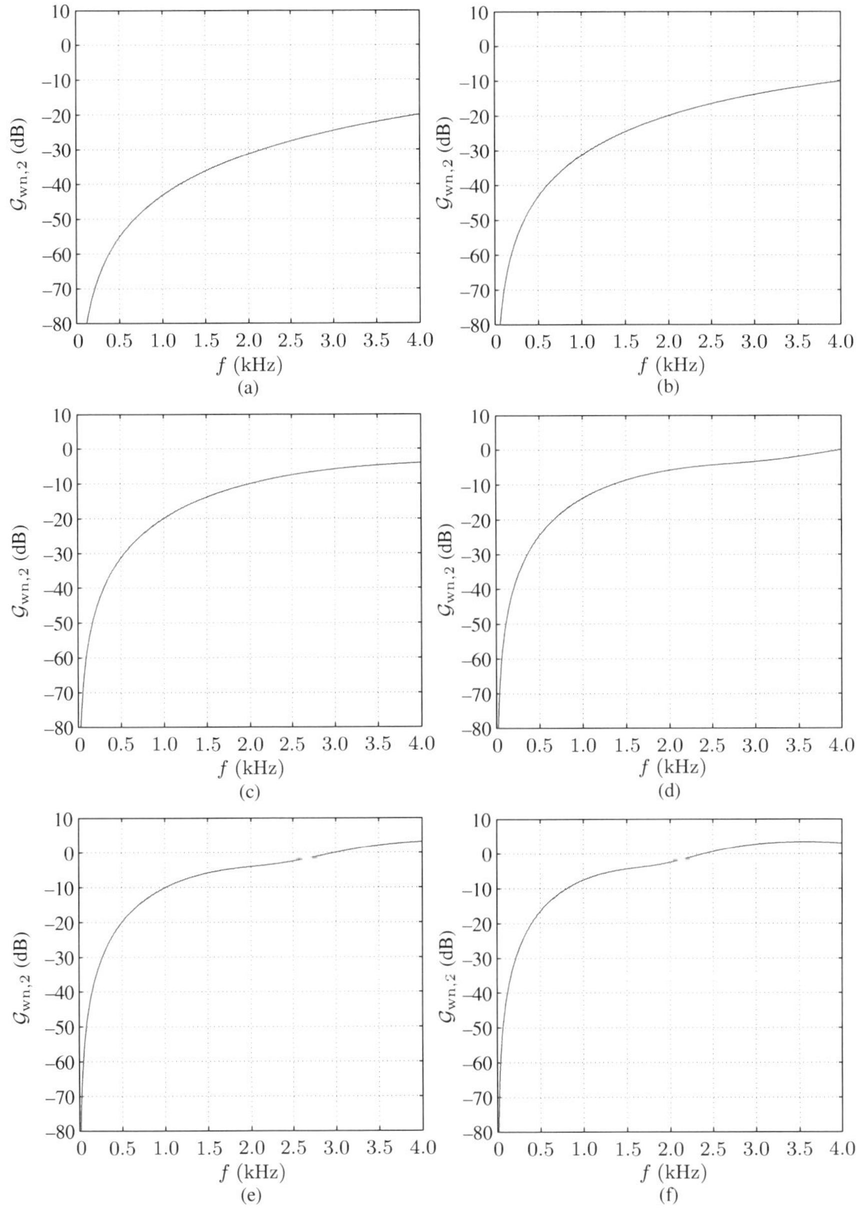

Fig. 7.6 The white noise gain of the second-order cardioid, as a function of frequency, for different values of δ: (a) $\delta = 0.5$ cm, (b) $\delta = 1$ cm, (c) $\delta = 2$ cm, (d) $\delta = 3$ cm, (e) $\delta = 4$ cm, and (f) $\delta = 5$ cm.

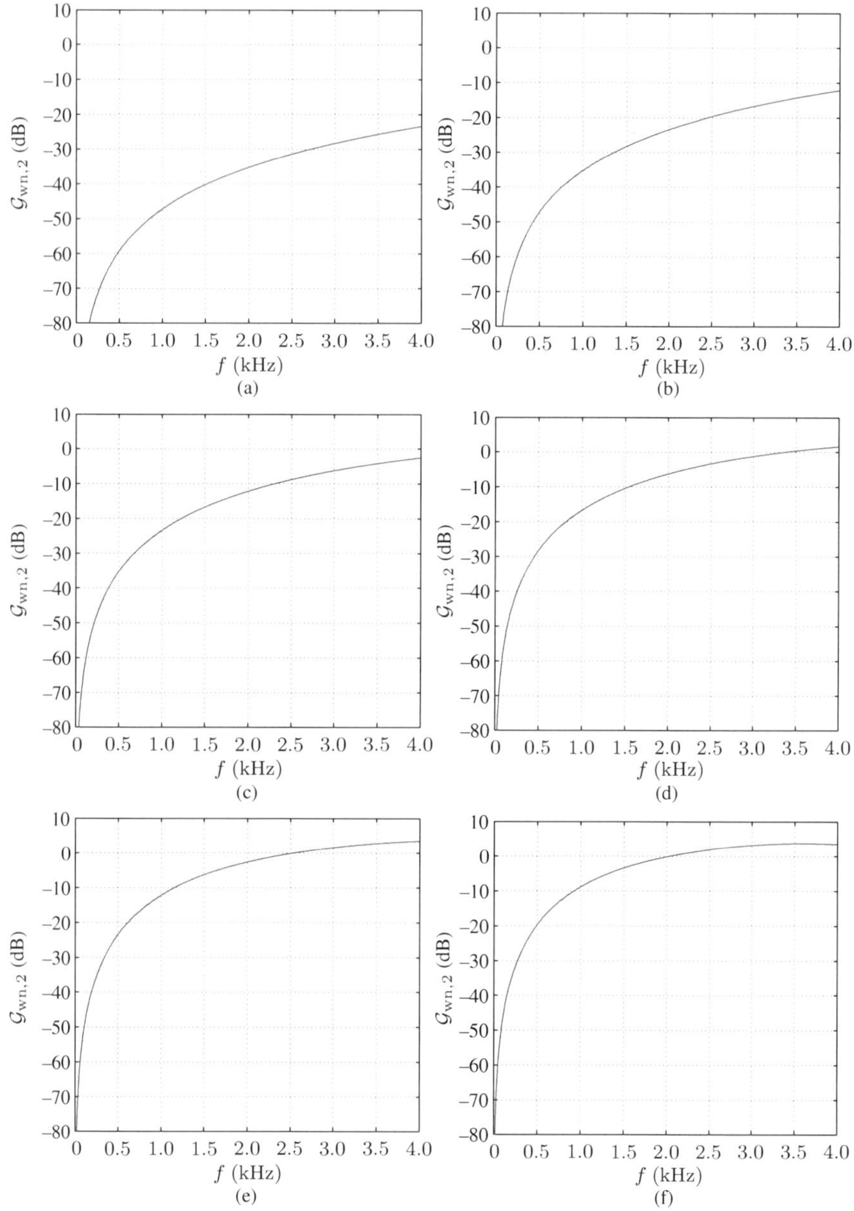

Fig. 7.7 The white noise gain of the second-order hypercardioid, as a function of frequency, for different values of δ: (a) $\delta = 0.5$ cm, (b) $\delta = 1$ cm, (c) $\delta = 2$ cm, (d) $\delta = 3$ cm, (e) $\delta = 4$ cm, and (f) $\delta = 5$ cm.

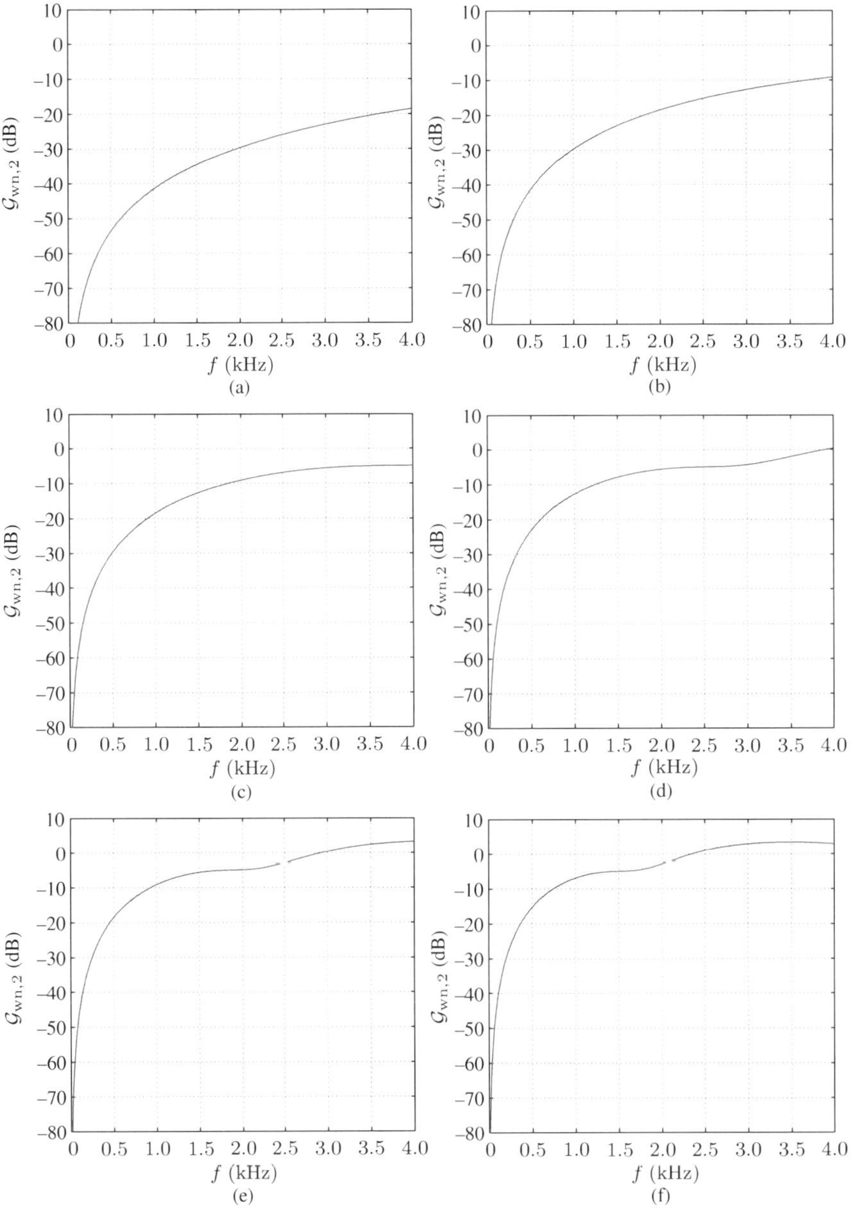

Fig. 7.8 The white noise gain of the second-order supercardioid, as a function of frequency, for different values of δ: (a) $\delta = 0.5$ cm, (b) $\delta = 1$ cm, (c) $\delta = 2$ cm, (d) $\delta = 3$ cm, (e) $\delta = 4$ cm, and (f) $\delta = 5$ cm.

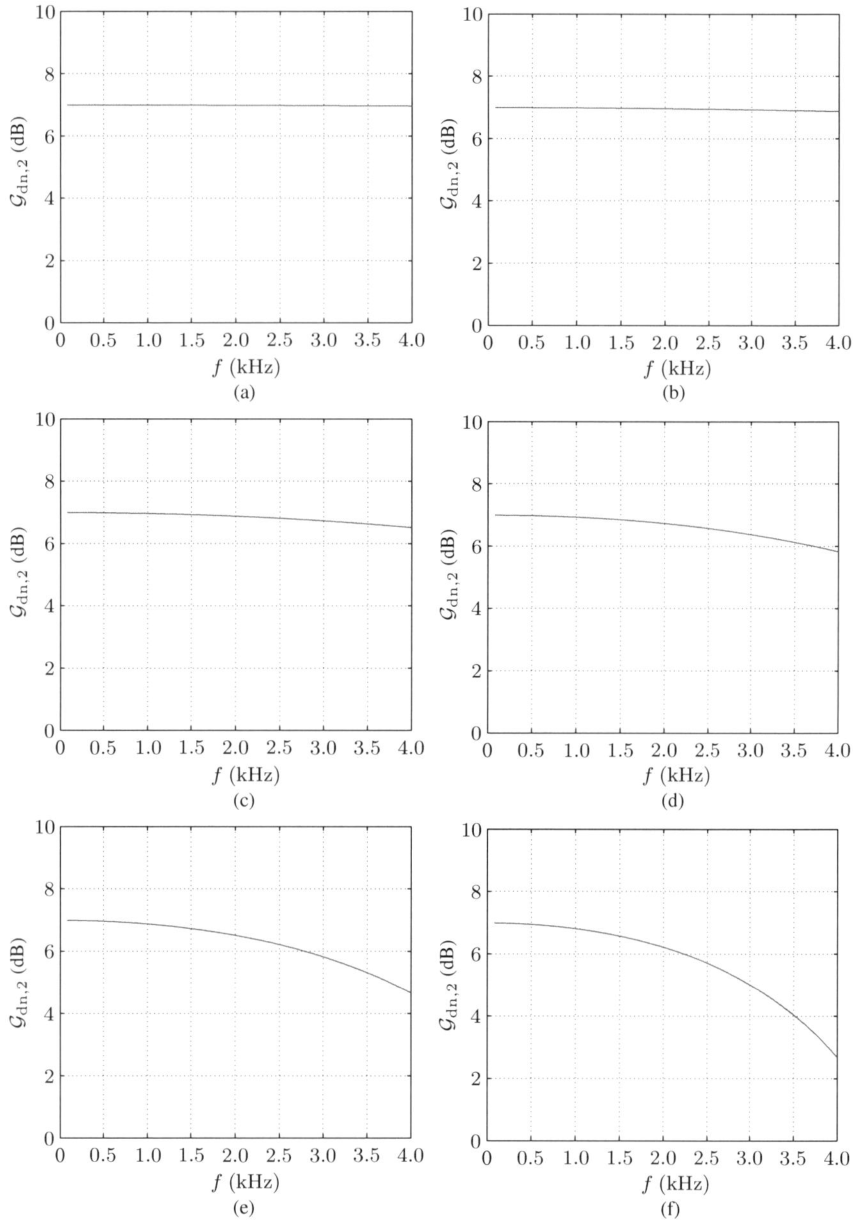

Fig. 7.9 The directivity factor of the second-order dipole, as a function of frequency, for different values of δ: (a) $\delta = 0.5$ cm, (b) $\delta = 1$ cm, (c) $\delta = 2$ cm, (d) $\delta = 3$ cm, (e) $\delta = 4$ cm, and (f) $\delta = 5$ cm.

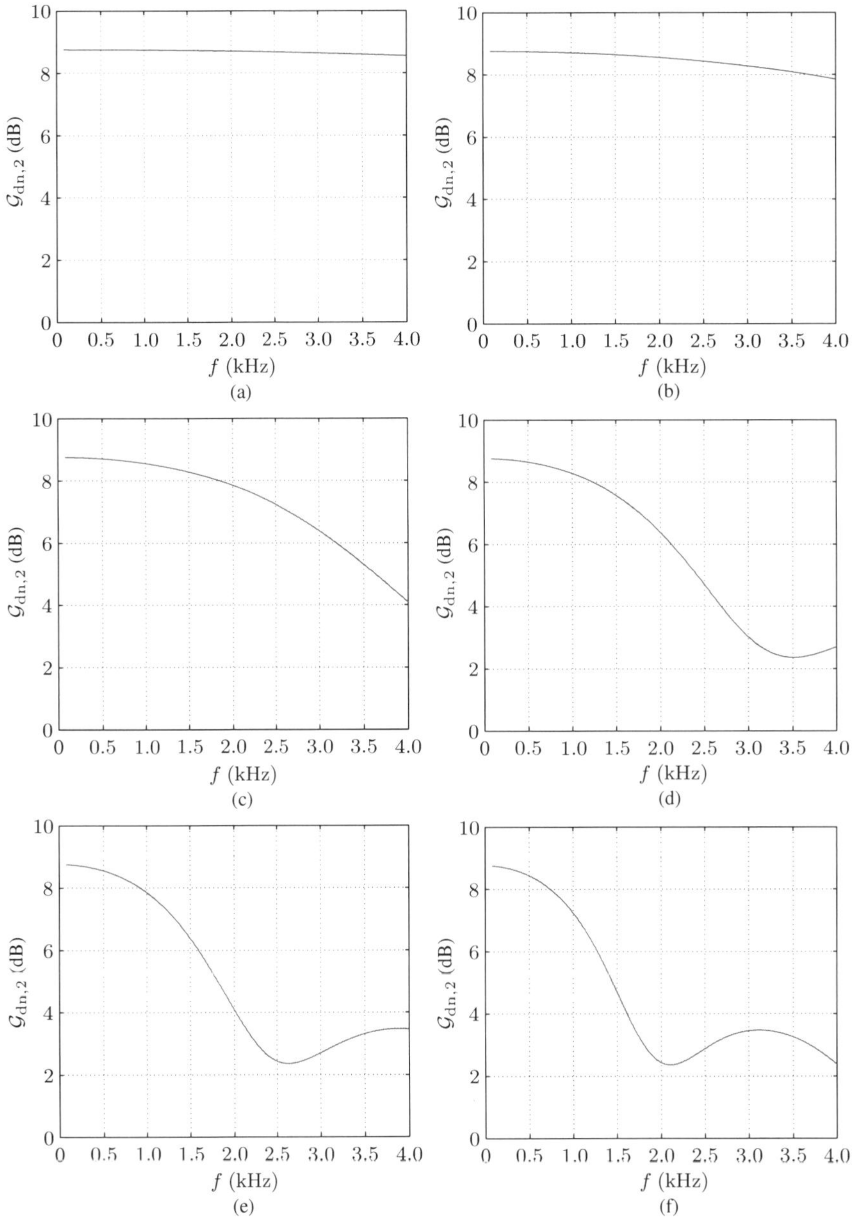

Fig. 7.10 The directivity factor of the second-order cardioid, as a function of frequency, for different values of δ: (a) $\delta = 0.5$ cm, (b) $\delta = 1$ cm, (c) $\delta = 2$ cm, (d) $\delta = 3$ cm, (e) $\delta = 4$ cm, and (f) $\delta = 5$ cm.

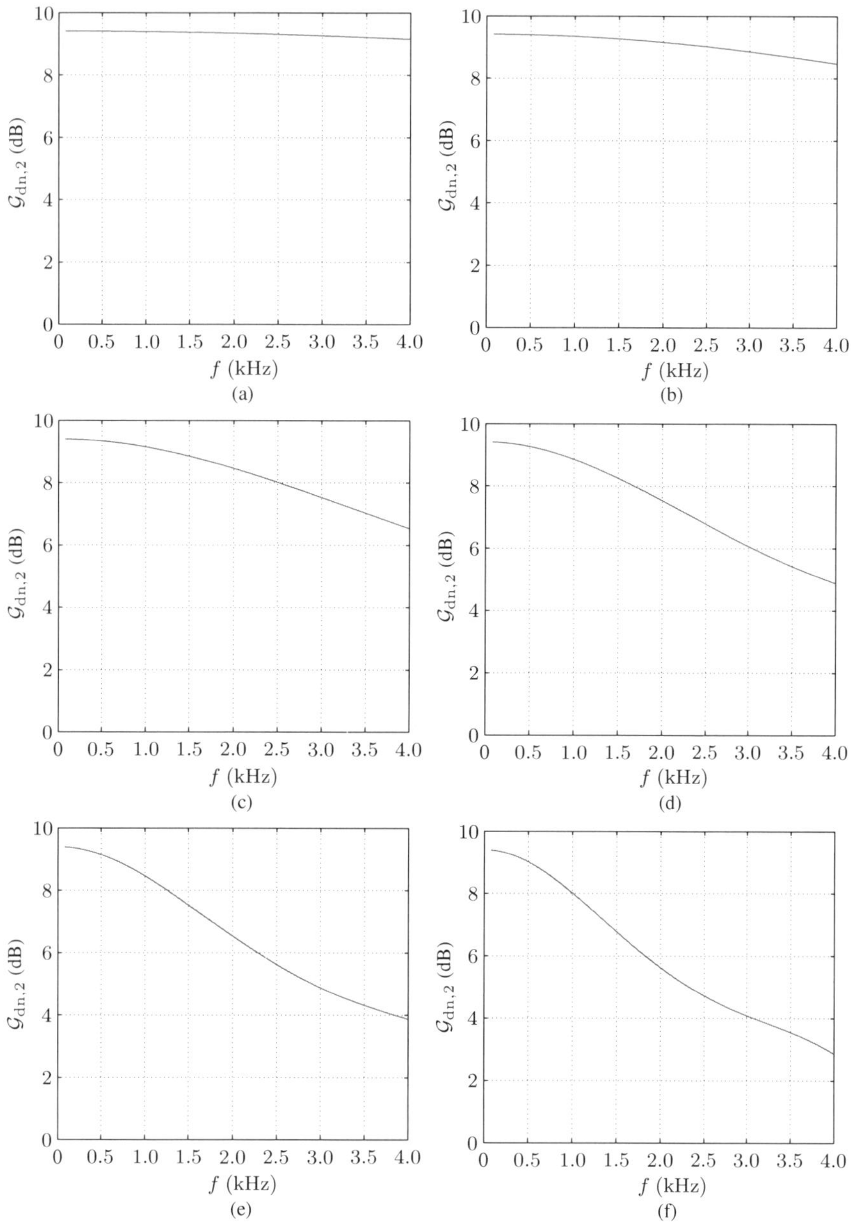

Fig. 7.11 The directivity factor of the second-order hypercardioid, as a function of frequency, for different values of δ: (a) $\delta = 0.5$ cm, (b) $\delta = 1$ cm, (c) $\delta = 2$ cm, (d) $\delta = 3$ cm, (e) $\delta = 4$ cm, and (f) $\delta = 5$ cm.

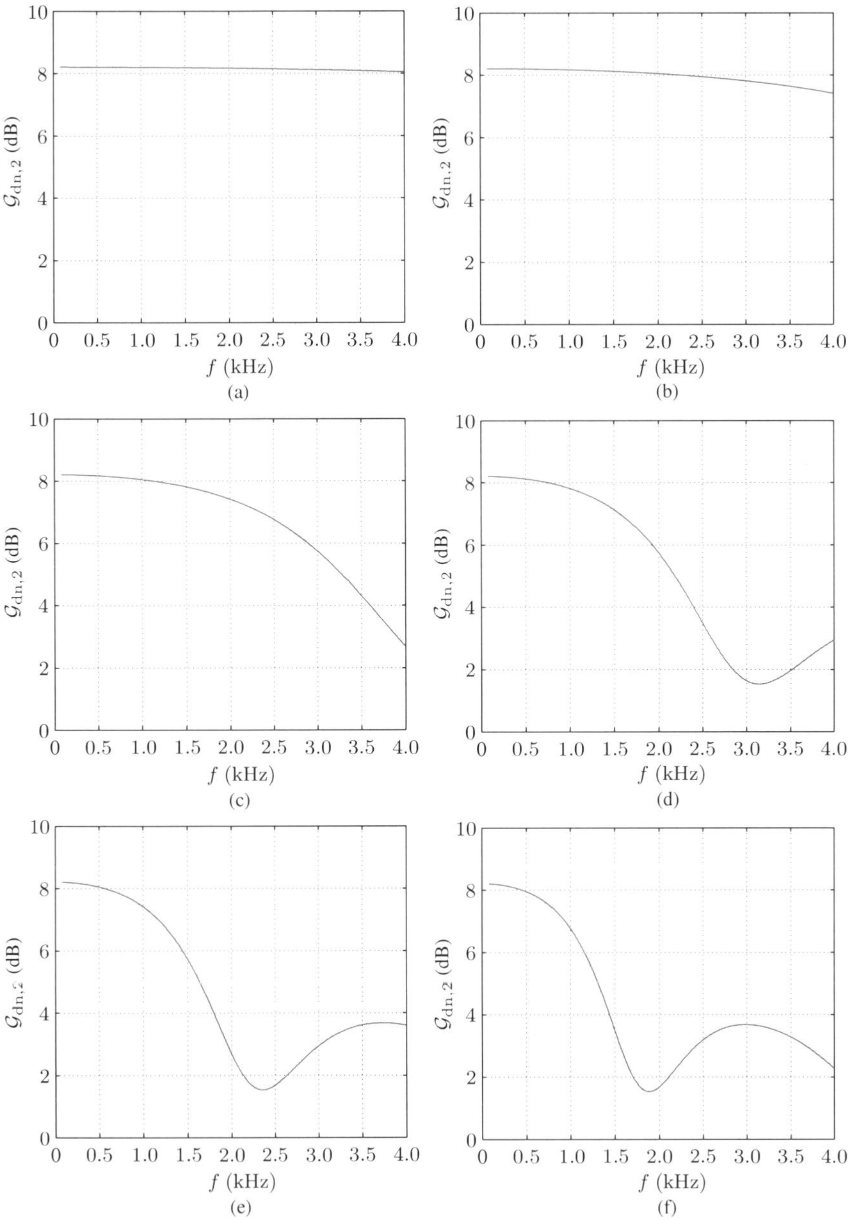

Fig. 7.12 The directivity factor of the second-order supercardioid, as a function of frequency, for different values of δ: (a) $\delta = 0.5$ cm, (b) $\delta = 1$ cm, (c) $\delta = 2$ cm, (d) $\delta = 3$ cm, (e) $\delta = 4$ cm, and (f) $\delta = 5$ cm.

7.4 Third-Order Differential Arrays

Following the same developments as in the previous two sections, we find that the inverse of $\mathbf{\Psi}_4$ is

$$\mathbf{\Psi}_4^{-1} = \begin{bmatrix} 1 & -\frac{11}{6} & 1 & -\frac{1}{6} \\ 0 & 3 & -\frac{5}{2} & \frac{1}{2} \\ 0 & -\frac{3}{2} & 2 & -\frac{1}{2} \\ 0 & \frac{1}{3} & -\frac{1}{2} & \frac{1}{6} \end{bmatrix}. \tag{7.33}$$

The filter is then

$$\mathbf{h}(\omega) = \left[H_1(\omega) \; H_2(\omega) \; H_3(\omega) \; H_4(\omega) \right]^T, \tag{7.34}$$

where

$$\begin{aligned} H_1(\omega) &= -\jmath \frac{a_{3,3}}{(\omega\tau_0)^3} \left\{ 1 - (\omega\tau_0)^2 \frac{11a_{3,1}}{6a_{3,3}} + \jmath \frac{(\omega\tau_0)^3}{a_{3,3}} \left[a_{3,0} - \frac{2a_{3,2}}{(\omega\tau_0)^2} \right] \right\}, \\ H_2(\omega) &= \jmath \frac{3a_{3,3}}{(\omega\tau_0)^3} \left[1 - \jmath\omega\tau_0 \frac{5a_{3,2}}{3a_{3,3}} - (\omega\tau_0)^2 \frac{a_{3,1}}{a_{3,3}} \right], \\ H_3(\omega) &= -\jmath \frac{3a_{3,3}}{(\omega\tau_0)^3} \left[1 - \jmath\omega\tau_0 \frac{4a_{3,2}}{3a_{3,3}} - (\omega\tau_0)^2 \frac{a_{3,1}}{2a_{3,3}} \right], \\ H_4(\omega) &= \jmath \frac{a_{3,3}}{(\omega\tau_0)^3} \left[1 - \jmath\omega\tau_0 \frac{a_{3,2}}{a_{3,3}} - (\omega\tau_0)^2 \frac{a_{3,1}}{3a_{3,3}} \right], \end{aligned}$$

from which we deduce the equivalent filter:

$$\mathbf{h}'(\omega) = \left[H_1'(\omega) \; H_2'(\omega) \; H_3'(\omega) \; H_4'(\omega) \right]^T, \tag{7.35}$$

where

$$\begin{aligned} H_1'(\omega) &= -\frac{1}{\omega^3} \left\{ 1 - (\omega\tau_0)^2 \frac{11a_{3,1}}{6a_{3,3}} + \jmath \frac{(\omega\tau_0)^3}{a_{3,3}} \left[a_{3,0} - \frac{2a_{3,2}}{(\omega\tau_0)^2} \right] \right\}, \\ H_2'(\omega) &= \frac{3}{\omega^3} \left[1 - \jmath\omega\tau_0 \frac{5a_{3,2}}{3a_{3,3}} - (\omega\tau_0)^2 \frac{a_{3,1}}{a_{3,3}} \right], \\ H_3'(\omega) &= -\frac{3}{\omega^3} \left[1 - \jmath\omega\tau_0 \frac{4a_{3,2}}{3a_{3,3}} - (\omega\tau_0)^2 \frac{a_{3,1}}{2a_{3,3}} \right], \\ H_4'(\omega) &= \frac{1}{\omega^3} \left[1 - \jmath\omega\tau_0 \frac{a_{3,2}}{a_{3,3}} - (\omega\tau_0)^2 \frac{a_{3,1}}{3a_{3,3}} \right]. \end{aligned}$$

The beampattern of the beamformer $\mathbf{h}'(\omega)$ is

$$\mathcal{B}\left[\mathbf{h}'(\omega),\theta\right] = \sum_{n=1}^{4} H'_n(\omega)\, e^{\jmath\omega(n-1)\tau_0\cos\theta}. \tag{7.36}$$

Figures 7.13, 7.14, 7.15, and 7.16 display the patterns from (7.36) for the dipole, cardioid, hypercardioid, and supercardioid, respectively, for several frequencies and two values of δ.

We are interested in gains of three types of noise:

- white,

$$\mathcal{G}_{\mathrm{wn},3}\left[\mathbf{h}'(\omega)\right] = \frac{\left|\sum_{n=1}^{4} H'_n(\omega)\, e^{\jmath\omega(n-1)\tau_0}\right|^2}{\sum_{n=1}^{4}\left|H'_n(\omega)\right|^2}, \tag{7.37}$$

- diffuse,

$$\mathcal{G}_{\mathrm{dn},3}\left[\mathbf{h}'(\omega)\right] = \frac{\left|\sum_{n=1}^{4} H'_n(\omega)\, e^{\jmath\omega(n-1)\tau_0}\right|^2}{\mathbf{h}'^H(\omega)\,\mathbf{\Gamma}_{\mathrm{dn}}(\omega)\,\mathbf{h}'(\omega)}, \tag{7.38}$$

- and point source,

$$\mathcal{G}_{\mathrm{ns},3}\left[\mathbf{h}'(\omega)\right] = \frac{\left|\sum_{n=1}^{4} H'_n(\omega)\, e^{\jmath\omega(n-1)\tau_0}\right|^2}{\left|\sum_{n=1}^{4} H'_n(\omega)\, e^{\jmath\omega(n-1)\tau_0\cos\theta_{\mathrm{n}}}\right|^2}. \tag{7.39}$$

In Figs. 7.17, 7.18, 7.19, and 7.20, we plot $\mathcal{G}_{\mathrm{wn},3}\left[\mathbf{h}'(\omega)\right]$ for the dipole, cardioid, hypercardioid, and supercardioid, respectively, for several frequencies and different values of δ.

In Figs. 7.21, 7.22, 7.23, and 7.24, we plot $\mathcal{G}_{\mathrm{dn},3}\left[\mathbf{h}'(\omega)\right]$ for the dipole, cardioid, hypercardioid, and supercardioid, respectively, for several frequencies and different values of δ.

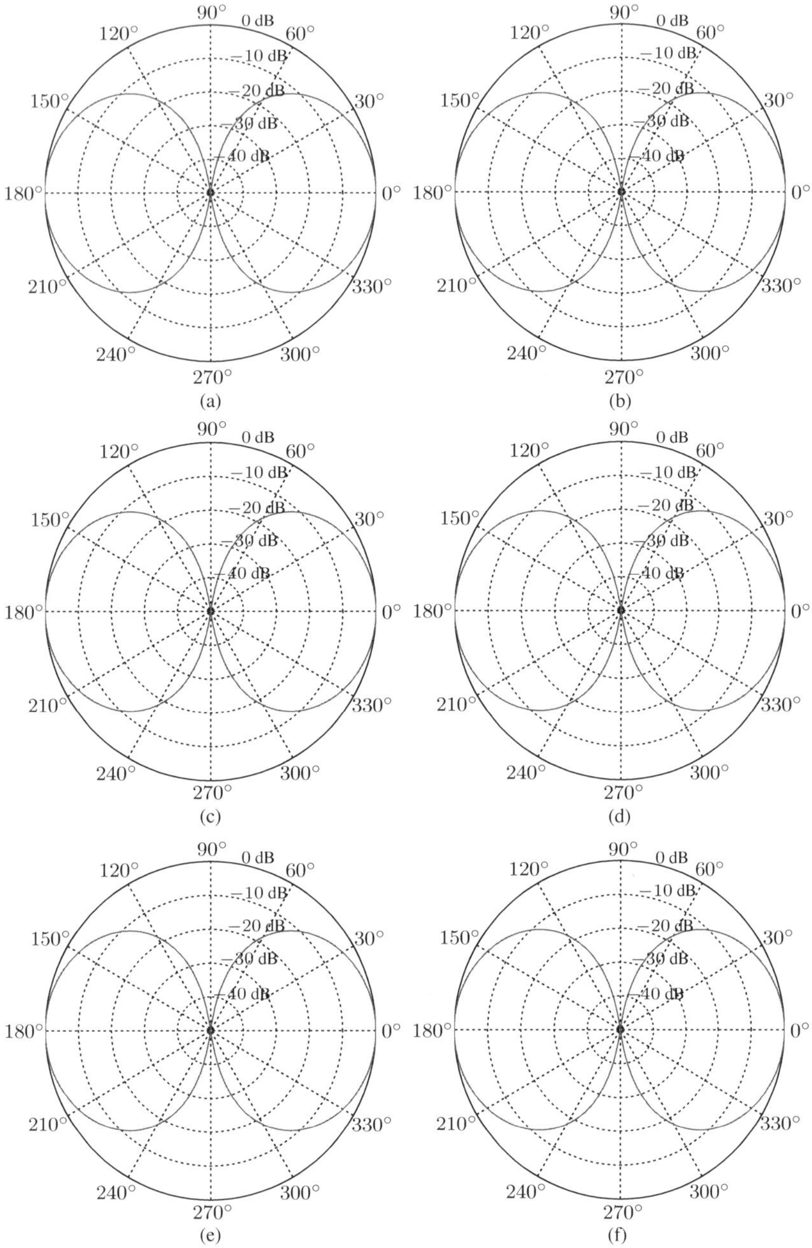

Fig. 7.13 Patterns of the third-order dipole for several frequencies and two values of δ: (a) $f = 0.5$ kHz, $\delta = 0.5$ cm, (b) $f = 0.5$ kHz, $\delta = 1$ cm, (c) $f = 1$ kHz, $\delta = 0.5$ cm, (d) $f = 1$ kHz, $\delta = 1$ cm, (e) $f = 3$ kHz, $\delta = 0.5$ cm, and (f) $f = 3$ kHz, $\delta = 1$ cm.

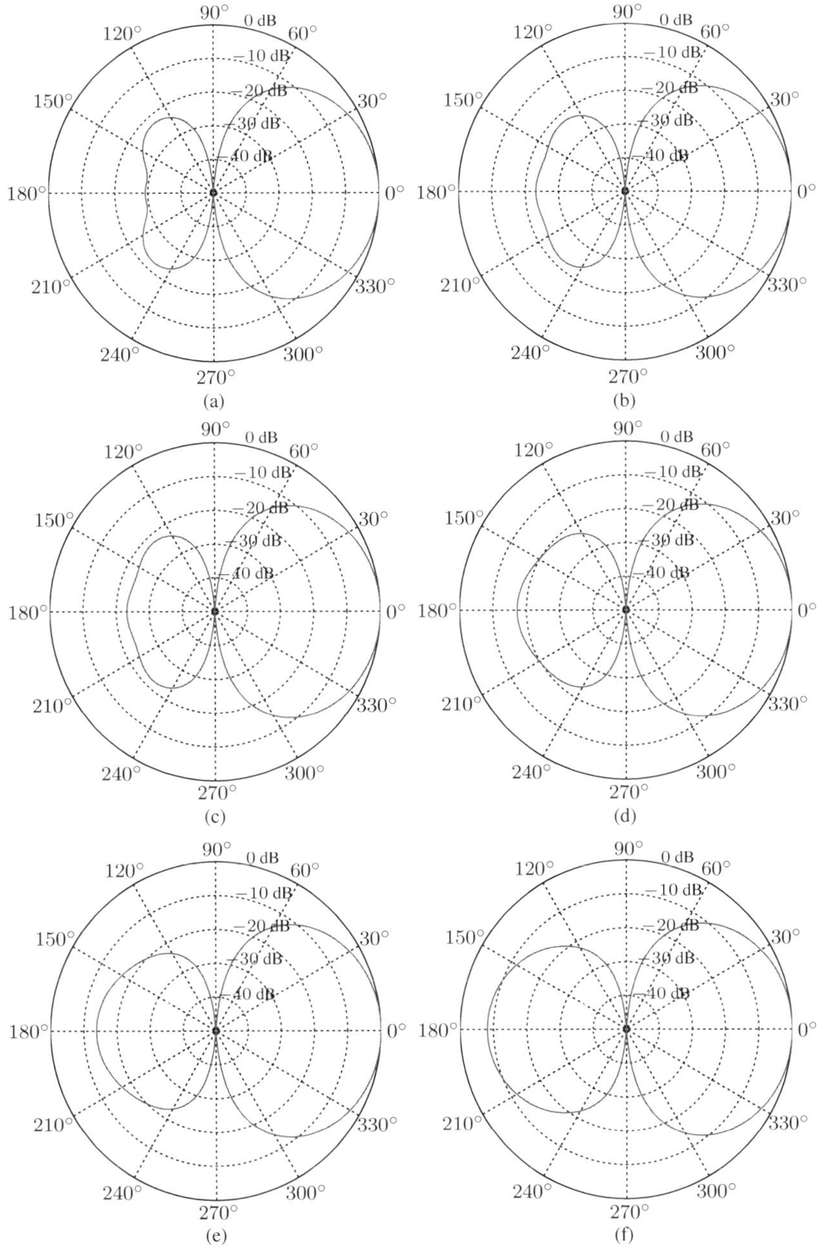

Fig. 7.14 Patterns of the third-order cardioid for several frequencies and two values of δ: (a) $f = 0.5$ kHz, $\delta = 0.5$ cm, (b) $f = 0.5$ kHz, $\delta = 1$ cm, (c) $f = 1$ kHz, $\delta = 0.5$ cm, (d) $f = 1$ kHz, $\delta = 1$ cm, (e) $f = 3$ kHz, $\delta = 0.5$ cm, and (f) $f = 3$ kHz, $\delta = 1$ cm.

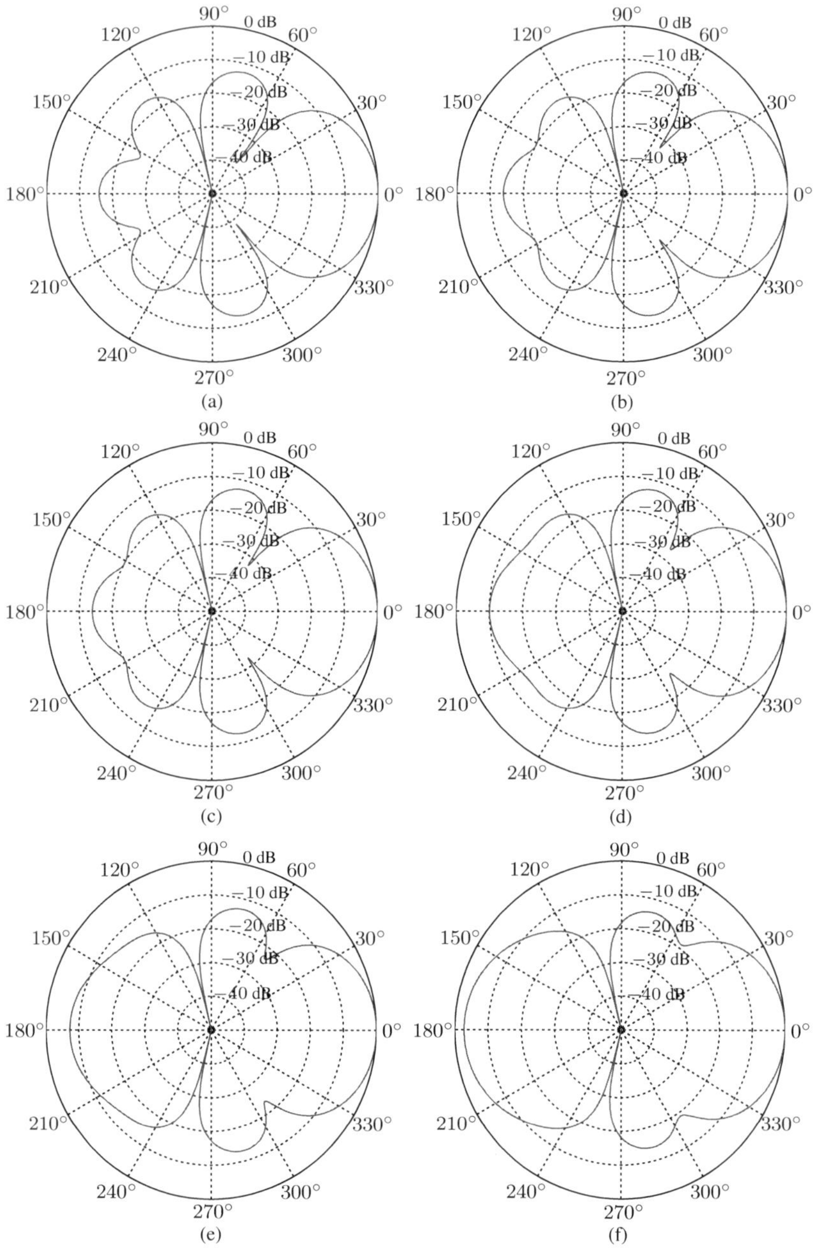

Fig. 7.15 Patterns of the third-order hypercardioid for several frequencies and two values of δ: (a) $f = 0.5$ kHz, $\delta = 0.5$ cm, (b) $f = 0.5$ kHz, $\delta = 1$ cm, (c) $f = 1$ kHz, $\delta = 0.5$ cm, (d) $f = 1$ kHz, $\delta = 1$ cm, (e) $f = 3$ kHz, $\delta = 0.5$ cm, and (f) $f = 3$ kHz, $\delta = 1$ cm.

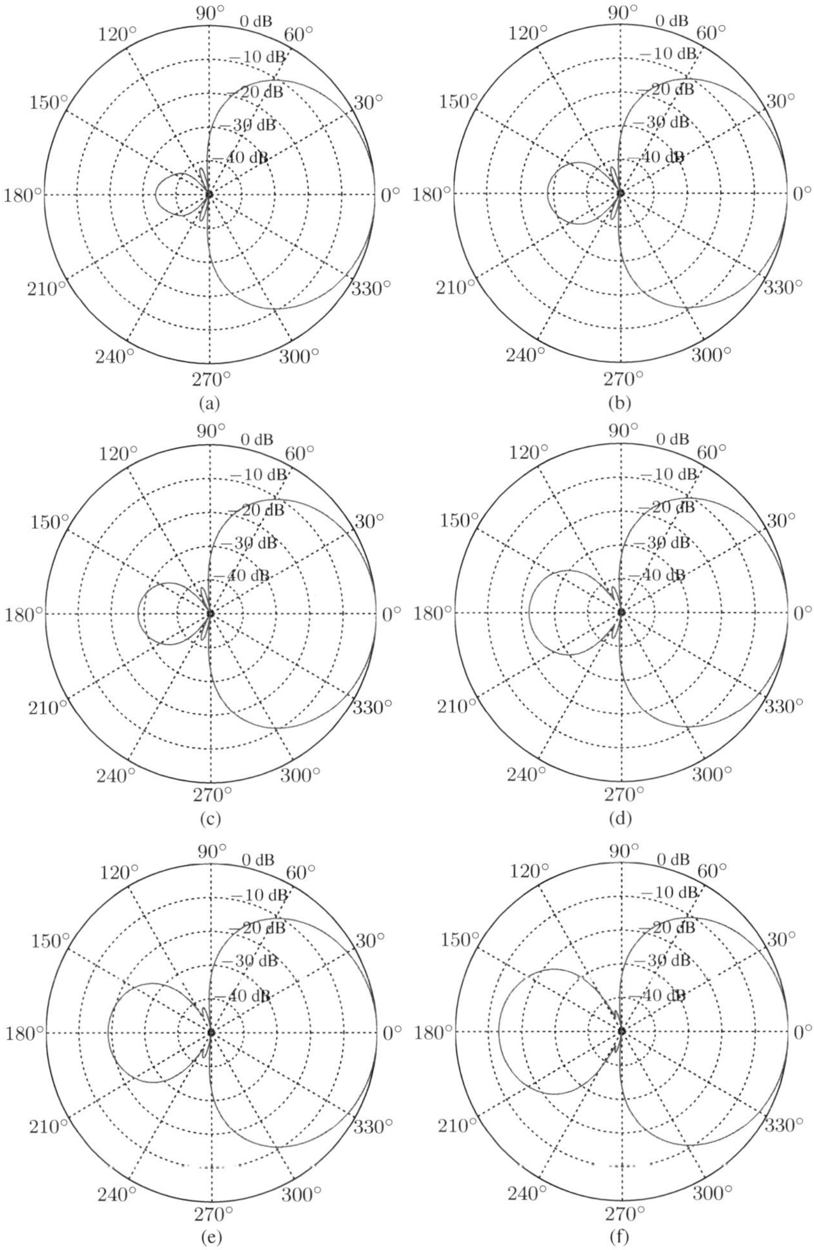

Fig. 7.16 Patterns of the third-order supercardioid for several frequencies and two values of δ: (a) $f = 0.5$ kHz, $\delta = 0.5$ cm, (b) $f = 0.5$ kHz, $\delta = 1$ cm, (c) $f = 1$ kHz, $\delta = 0.5$ cm, (d) $f = 1$ kHz, $\delta = 1$ cm, (e) $f = 3$ kHz, $\delta = 0.5$ cm, and (f) $f = 3$ kHz, $\delta = 1$ cm.

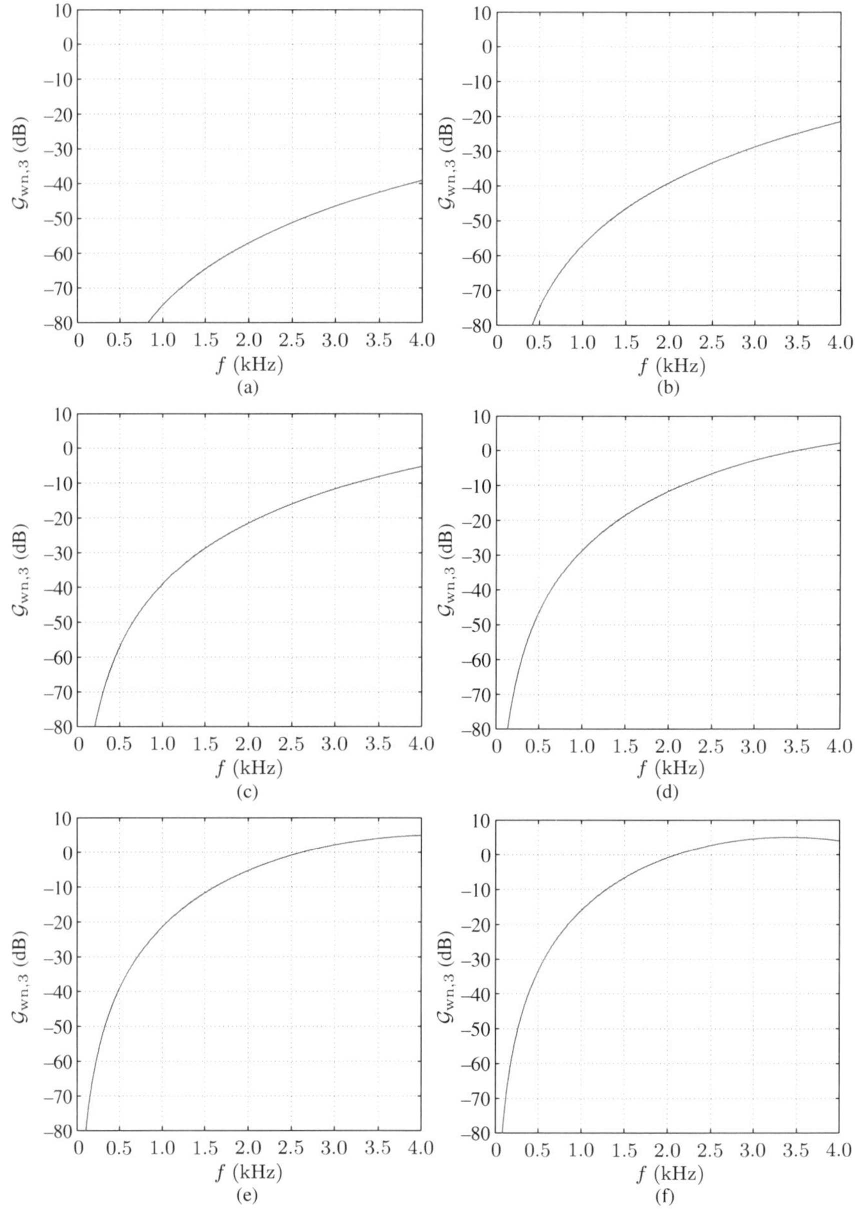

Fig. 7.17 The white noise gain of the third-order dipole, as a function of frequency, for different values of δ: (a) $\delta = 0.5$ cm, (b) $\delta = 1$ cm, (c) $\delta = 2$ cm, (d) $\delta = 3$ cm, (e) $\delta = 4$ cm, and (f) $\delta = 5$ cm.

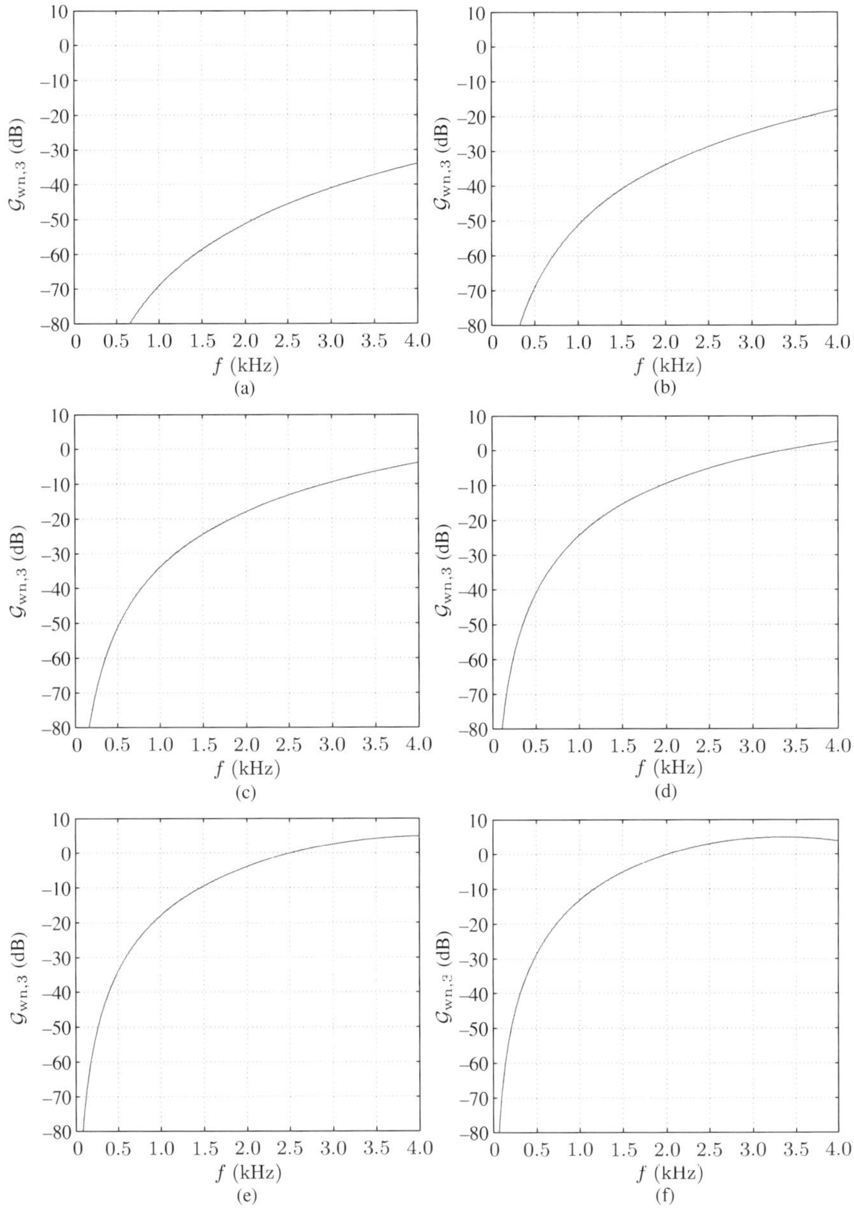

Fig. 7.18 The white noise gain of the third-order cardioid, as a function of frequency, for different values of δ: (a) $\delta = 0.5$ cm, (b) $\delta = 1$ cm, (c) $\delta = 2$ cm, (d) $\delta = 3$ cm, (e) $\delta = 4$ cm, and (f) $\delta = 5$ cm.

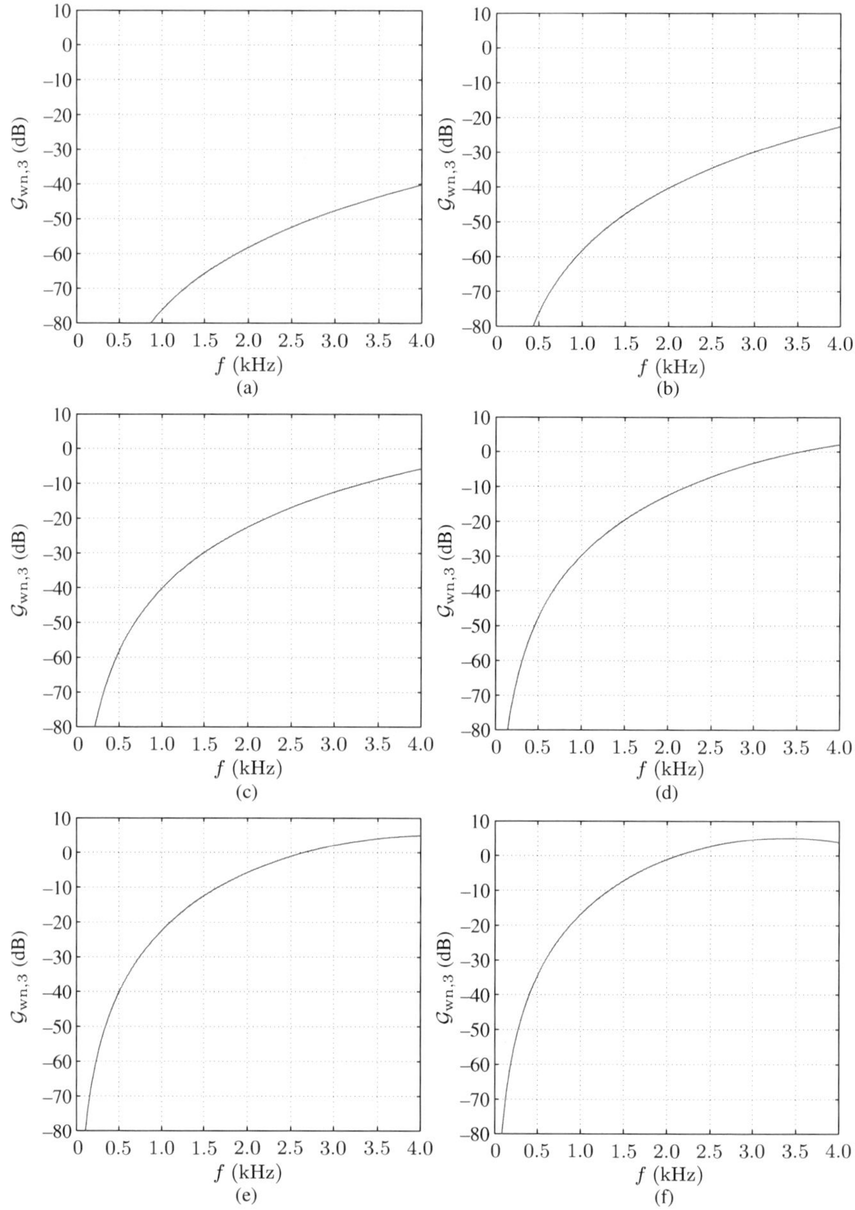

Fig. 7.19 The white noise gain of the third-order hypercardioid, as a function of frequency, for different values of δ: (a) $\delta = 0.5$ cm, (b) $\delta = 1$ cm, (c) $\delta = 2$ cm, (d) $\delta = 3$ cm, (e) $\delta = 4$ cm, and (f) $\delta = 5$ cm.

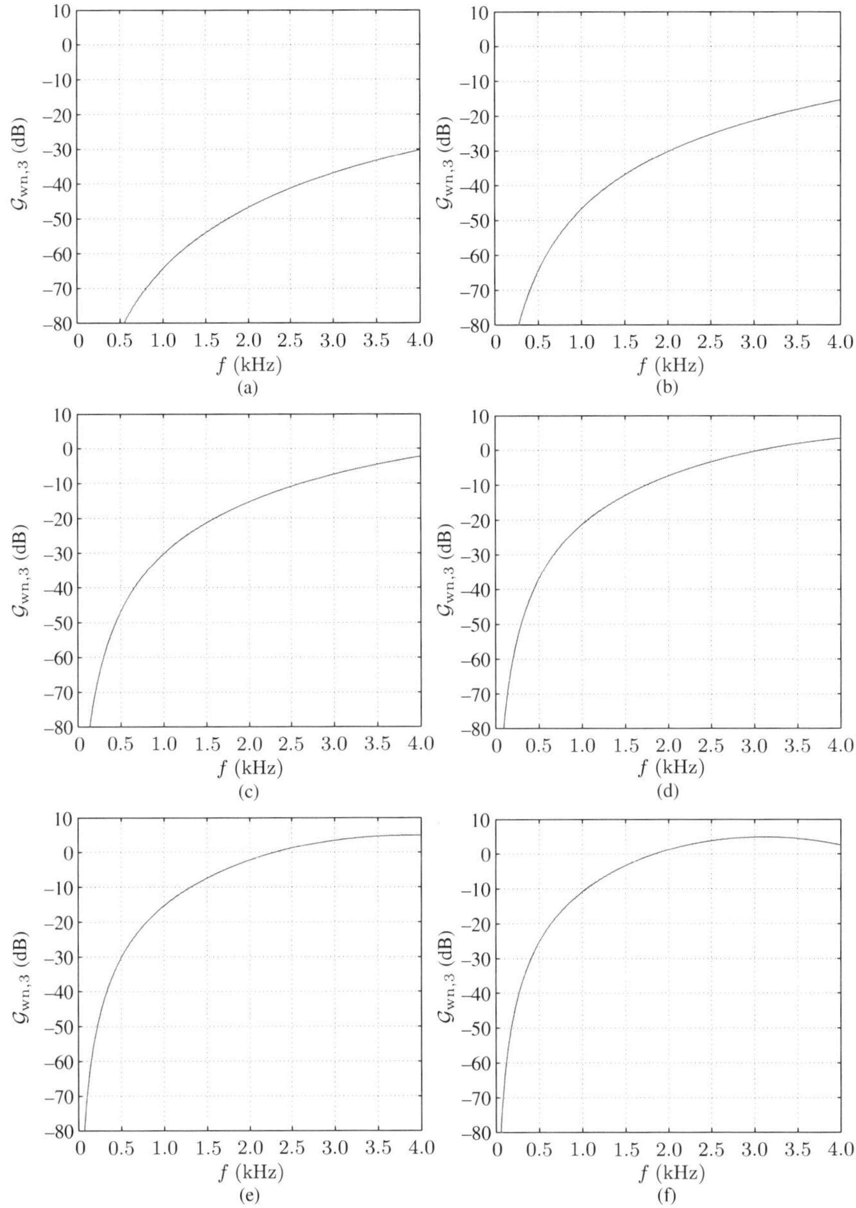

Fig. 7.20 The white noise gain of the third-order supercardioid, as a function of frequency, for different values of δ: (a) $\delta = 0.5$ cm, (b) $\delta = 1$ cm, (c) $\delta = 2$ cm, (d) $\delta = 3$ cm, (e) $\delta = 4$ cm, and (f) $\delta = 5$ cm.

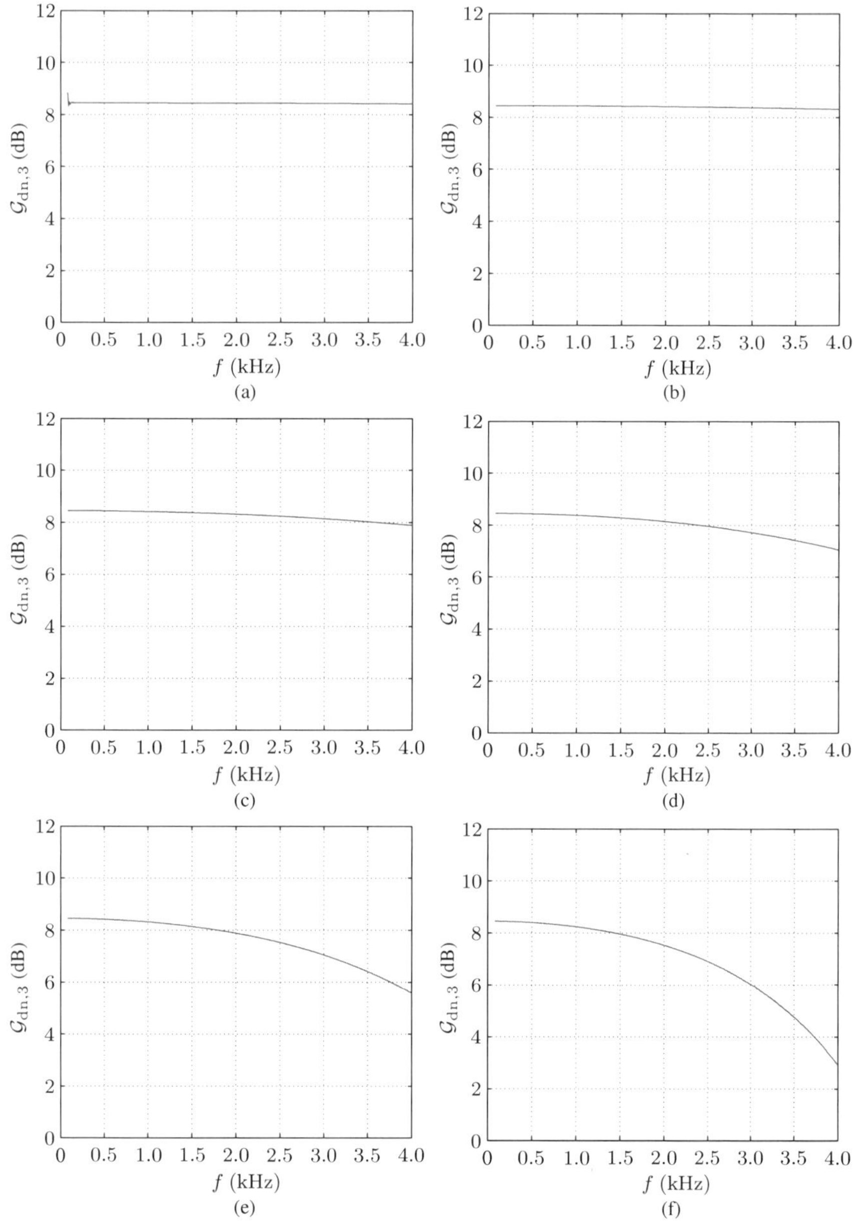

Fig. 7.21 The directivity factor of the third-order dipole, as a function of frequency, for different values of δ: (a) $\delta = 0.5$ cm, (b) $\delta = 1$ cm, (c) $\delta = 2$ cm, (d) $\delta = 3$ cm, (e) $\delta = 4$ cm, and (f) $\delta = 5$ cm.

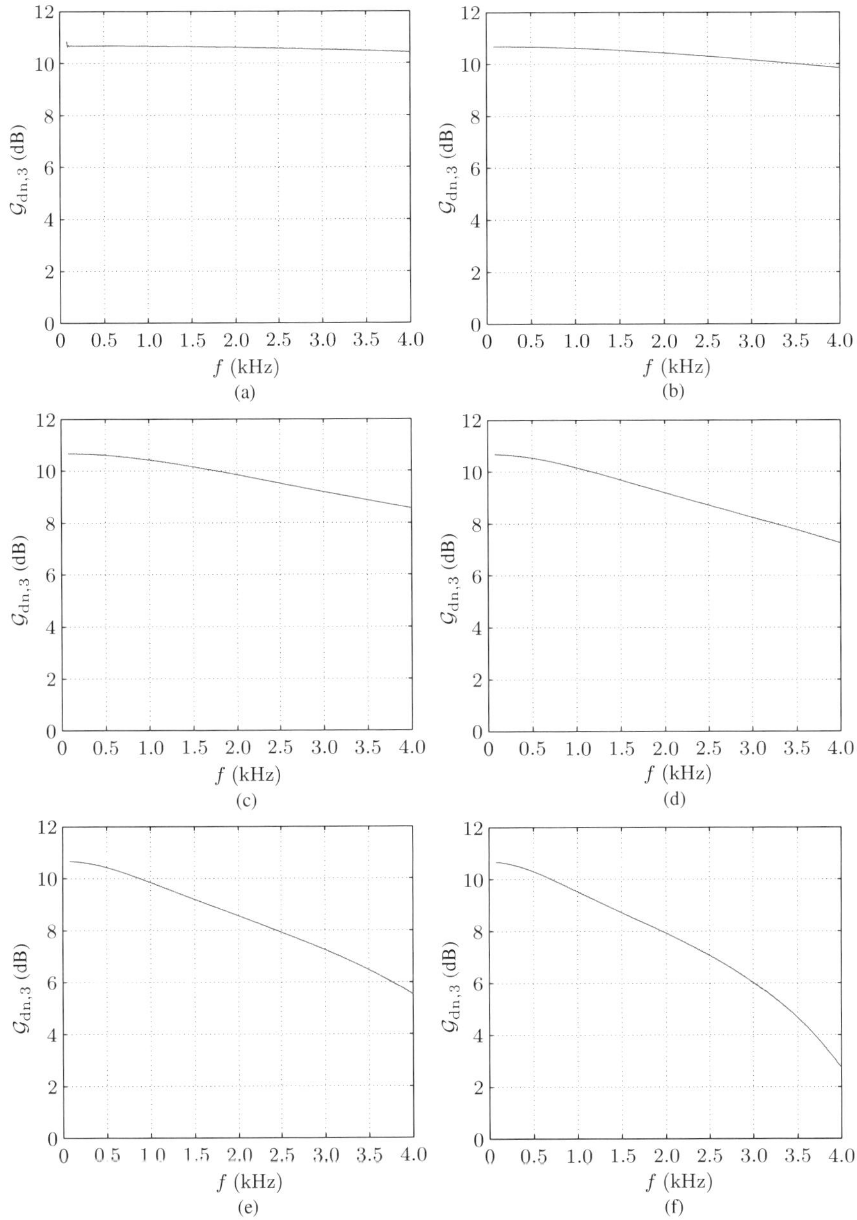

Fig. 7.22 The directivity factor of the third-order cardioid, as a function of frequency, for different values of δ: (a) $\delta = 0.5$ cm, (b) $\delta = 1$ cm, (c) $\delta = 2$ cm, (d) $\delta = 3$ cm, (e) $\delta = 4$ cm, and (f) $\delta = 5$ cm.

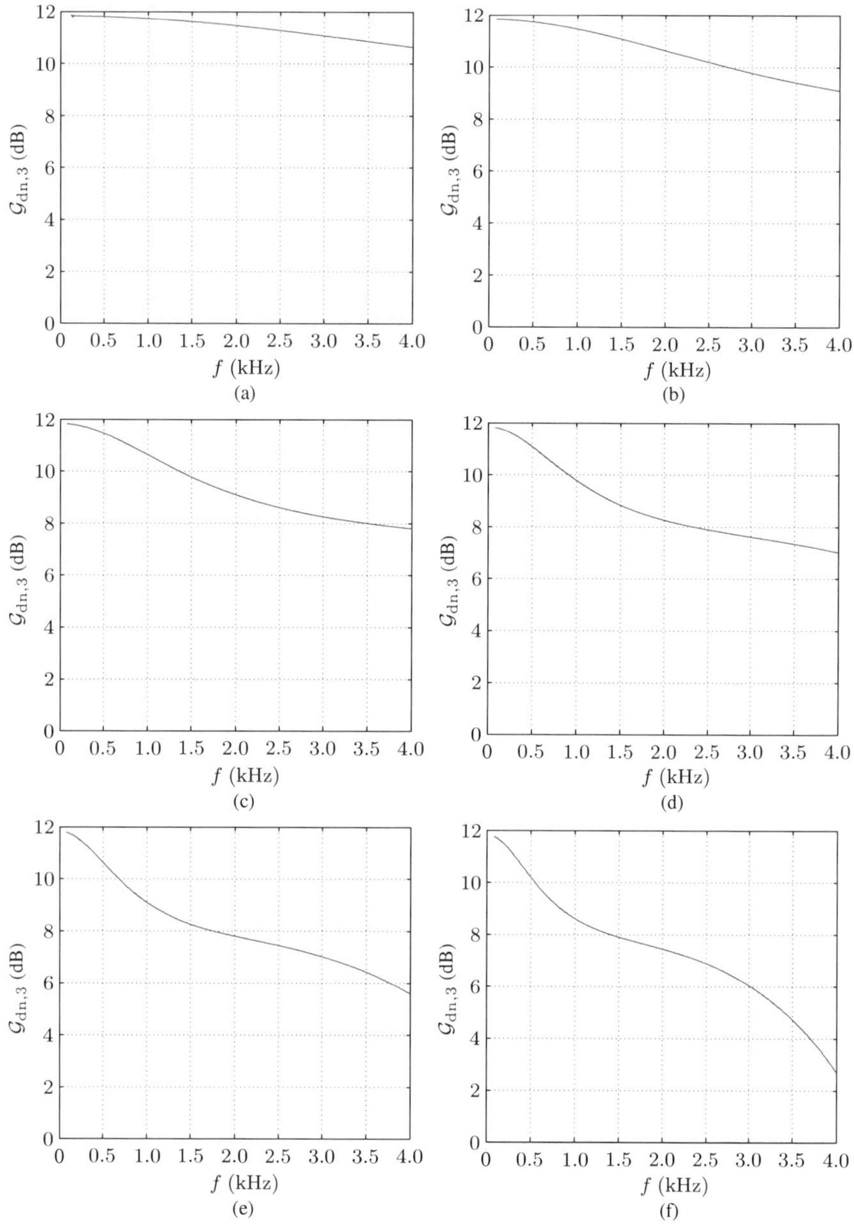

Fig. 7.23 The directivity factor of the third-order hypercardioid, as a function of frequency, for different values of δ: (a) $\delta = 0.5$ cm, (b) $\delta = 1$ cm, (c) $\delta = 2$ cm, (d) $\delta = 3$ cm, (e) $\delta = 4$ cm, and (f) $\delta = 5$ cm.

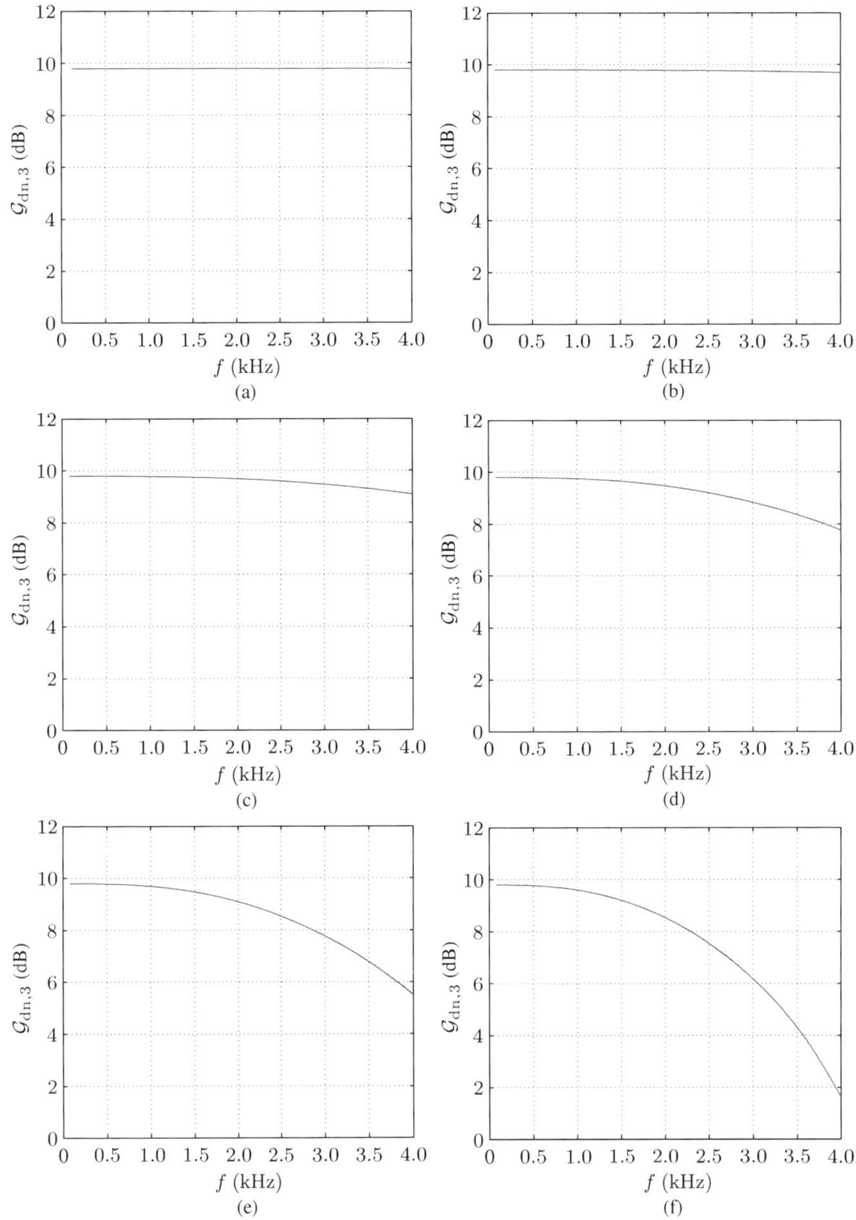

Fig. 7.24 The directivity factor of the third-order supercardioid, as a function of frequency, for different values of δ: (a) $\delta = 0.5$ cm, (b) $\delta = 1$ cm, (c) $\delta = 2$ cm, (d) $\delta = 3$ cm, (e) $\delta = 4$ cm, and (f) $\delta = 5$ cm.

7.5 Adaptive Beamforming versus Differential Arrays

The linear system given in (7.3) can be generalized to

$$\mathbf{\Psi}_{N+1}\mathbf{h}(\omega) = \mathbf{a}_{N+1}(\omega), \tag{7.40}$$

where now

$$\mathbf{\Psi}_{N+1} = \begin{bmatrix} 1 & 1 & 1 & \cdots & 1 \\ 0 & 1 & 2 & \cdots & M-1 \\ 0 & 1 & 2^2 & \cdots & (M-1)^2 \\ \vdots & \vdots & \vdots & \ddots & \vdots \\ 0 & 1 & 2^N & \cdots & (M-1)^N \end{bmatrix} \tag{7.41}$$

is the constraint matrix of size $(N+1)\times M$ having the Vandermonde structure, M is the number of microphones,

$$\mathbf{h}(\omega) = \left[H_1(\omega)\; H_2(\omega) \cdots H_M(\omega) \right]^T \tag{7.42}$$

is a filter of length M, and the vector $\mathbf{a}_{N+1}(\omega)$ was already defined in (7.5).

In adaptive beamforming [2], we minimize the residual noise at the beamformer output subject to the constraints summarized in (7.40). Mathematically, this is equivalent to

$$\min_{\mathbf{h}(\omega)} \mathbf{h}^H(\omega)\mathbf{\Phi}_{\mathbf{v}}(\omega)\mathbf{h}(\omega) \quad \text{subject to} \quad \mathbf{\Psi}_{N+1}\mathbf{h}(\omega) = \mathbf{a}_{N+1}(\omega). \tag{7.43}$$

We easily deduce that the solution is the LCMV filter [2], [3]:

$$\mathbf{h}_{\text{LCMV}}(\omega) = \mathbf{\Phi}_{\mathbf{v}}^{-1}(\omega)\mathbf{\Psi}_{N+1}^T\left[\mathbf{\Psi}_{N+1}\mathbf{\Phi}_{\mathbf{v}}^{-1}(\omega)\mathbf{\Psi}_{N+1}^T\right]^{-1}\mathbf{a}_{N+1}(\omega). \tag{7.44}$$

We observe that for the matrix $\mathbf{\Psi}_{N+1}\mathbf{\Phi}_{\mathbf{v}}^{-1}(\omega)\mathbf{\Psi}_{N+1}^T$ in (7.44) to be full rank, we must have $N+1 \leq M$, which is the same condition to design a differential array of order N.

For $M = N+1$, we easily deduce from (7.44) that

$$\mathbf{h}_{\text{LCMV}}(\omega) = \mathbf{\Psi}_{N+1}^{-1}\mathbf{a}_{N+1}(\omega), \tag{7.45}$$

which corresponds exactly to the filter of an Nth-order DMA or the solution of (7.40).

For $M > N+1$ and spatially white noise, (7.44) becomes

$$\mathbf{h}_{\text{LCMV}}(\omega) = \mathbf{\Psi}_{N+1}^T\left(\mathbf{\Psi}_{N+1}\mathbf{\Psi}_{N+1}^T\right)^{-1}\mathbf{a}_{N+1}(\omega), \tag{7.46}$$

which corresponds to the minimum-norm solution of (7.40). This shows that the LCMV filter is fundamentally related to the filter of an Nth-order DMA.

The minimum-norm filter given in (7.46) can also be a good solution against white noise amplification.

References

1. G. W. Elko, "Superdirectional microphone arrays," in *Acoustic Signal Processing for Telecommunication*, S. L. Gay and J. Benesty, Eds. Boston, MA: Kluwer Academic Publishers, 2000, Chapter 10, pp. 181–237.
2. J. Benesty, J. Chen, and Y. Huang, *Microphone Array Signal Processing*. Berlin, Germany: Springer-Verlag, 2008.
3. O. Frost, "An algorithm for linearly constrained adaptive array processing," *Proc. IEEE*, vol. 60, pp. 926–935, Jan. 1972.

Index

J. Benesty and J. Chen, *Study and Design of Differential Microphone Arrays*,
Springer Topics in Signal Processing 6, DOI: 10.1007/978-3-642-33753-6,

Printed by Printforce, the Netherlands